Early, Rapid and Sensitive Veterinary Molecular Diagnostics - Real Time PCR Applications

Ericka A. Pestana · Sandor Belak · Adama Diallo · John R. Crowther · Gerrit J. Viljoen

Early, Rapid and Sensitive Veterinary Molecular Diagnostics - Real Time PCR Applications

Springer

Dr. Ericka A. Pestana
Department of Nuclear Sciences
and Applications
International Atomic Energy
Agency (IAEA)
A1400 Vienna, Austria
E.A.Pestana-Delgado@iaea.org

Prof. Sandor Belak
Swedish University of
Agricultural Sciences
Department of Virology
SE-750 07 Uppsala
Sweden
sandor.belak@bvf.slu.se

Dr. Adama Diallo
Department of Nuclear Sciences
and Applications
International Atomic Energy
Agency (IAEA)
A1400 Vienna, Austria
adama.diallo@iaea.org

Dr. John R. Crowther
Department of Nuclear Sciences
and Applications
International Atomic Energy
Agency (IAEA)
A1400 Vienna, Austria
j.crowther@iaea.org

Prof. Gerrit J. Viljoen
Department of Nuclear Sciences
and Applications
International Atomic Energy
Agency (IAEA)
A1400 Vienna, Austria
G.J.Viljoen@iaea.org

ISBN 978-90-481-3131-0 e-ISBN 978-90-481-3132-7
DOI 10.1007/978-90-481-3132-7
Springer Dordrecht Heidelberg London New York

Library of Congress Control Number: 2009938825

Printed on acid-free paper

Springer is part of Springer Science+Business Media (www.springer.com)

Foreword

The Joint FAO/IAEA Division of Nuclear Techniques in Food and Agriculture is involved in agricultural research and development and assists Member States of FAO and IAEA in improving strategies to ensure food security through the use of nuclear techniques and related biotechnologies, where such techniques have a valuable and often unique role. In particular, nuclear and nuclear related molecular diagnostic methods have rapidly evolved in the past 20 years, since the advent of the Polymerase Chain Reaction (PCR). They are used in a wide range of agricultural areas such as, improving soil and water management; producing better crop varieties; diagnosing plant and animal diseases; controlling insect pests and improving food quality and safety.

The uses of nucleic acid-directed methods have increased significantly in the past 5 years and have made important contributions to disease control country programmes for improving national and international trade. These developments include the more routine use of PCR, Real-Time PCR and PCR-Sequencing as diagnostic and characterization tools in veterinary diagnostic laboratories. However, there are many problems associated with the transfer and particularly, the application of this technology. These include lack of consideration of the establishment of quality-assured procedures, the required set-up of the laboratory and the proper training of staff. This can lead to a situation where results are not assured.

This book gives a comprehensive account of the practical aspects of real-time PCR and strong consideration is given to ensure its optimal use in a diagnostic laboratory environment. This includes the basic principles, setting-up of a Real-Time PCR laboratory; Good Laboratory Practice and Standard Operating Procedures; Diagnostic Implementation, Execution and Interpretation, Analysis and Problem Solving. Examples of Standard Operating Procedures as used in individual specialist laboratories and an outline of training materials necessary for Real-Time PCR technology transfer are presented. The difficulties, advantages and disadvantages in PCR and Real-Time PCR applications are explained and placed in context with other test systems.

Emphasis is placed on the use of Real-Time PCR for detection of pathogens, with a particular focus on diagnosticians and scientists from the developing world.

It is hoped that this book will enable readers from various disciplines and levels of expertise to better judge the merits of early and rapid nuclear and nuclear related molecular diagnostic approaches and to increase their skills and knowledge in order to assist in a more logical, efficient and assured use of these technologies.

Liang Qu
Director:
Joint FAO/IAEA Division of Nuclear Techniques in Food and Agriculture
Department of Nuclear Sciences and Applications

Acknowledgements

The authors would like to thank the following people and Institutions for their contributions, suggestions and permissions:

Donald King and the Institute of Animal Health, Pirbright Laboratory, United Kingdom; Sharon Hietala and the California Animal Health and Food Safety Laboratory; Ian Brown and the Veterinary Laboratories Agency, Surrey-United Kingdom; Janusz Paweska and the Special Pathogens Unit (NICD-NHLS) in Sandringham-Johannesburg, for their valuable contributions on specific procedures for disease diagnostics. Bio-Rad, Roche Diagnostics GmbH; PE Biosystems; Applied Biosystems; Perkin Elmer, and Stratagene; for their "visual" help.

Thanks to the people at the SVA-SLU Virology R&D Division, at the MOD section of SVA VIP and at Uppsala University (the groups of Professors Jonas Blomberg and Ulf Landegren).

Many thanks to Anne-Lie Blomstram et al. and Lihong Liu et al. (see refs) for providing figures.

The authors also would like to acknowledge SVA and SLU, FORMAS, AGRIA, KBM, as well as the European FP6 and FP7 projects and their research consortia: LAB-ON-SITE (SSP3-513Â 645), CSF&WILD-BOAR (SSP1-501599), EPIZONE (FOOD-CT-2006-016236), FLUTEST and FLUTRAIN.

Special thanks to Marcus Neusser from Bio-Rad, for his advice and help on completing this venture, and to all those who peer reviewed this book.

Contents

Chapter 1
Background

Contents

1.1 Aims of This Book

The Joint FAO/IAEA Division of Nuclear Techniques in Food and Agriculture (NAFA) and Nuclear Sciences and Applications Laboratories (NAAL) are jointly involved in research and development in the fields of agriculture with particular responsibility to the diagnostic laboratories of their Member States. Nuclear related techniques play an increasingly valuable and often unique role in agricultural research and development. They have been applied in a wide range of disciplines including improving soil and water management; producing better crop varieties; diagnosing animal diseases; controlling insect pests and improving food quality and safety. The temperature template amplification technologies - Polymerase Chain Reaction (PCR), with their already proven value and massive potential, are at the forefront of debate and interest to a wide range of scientists in developing and developed countries alike. It is relevant to note that real-time PCR (rt-PCR) and quantitative PCR (qPCR) are derivatives of the polymerase chain reaction. This book extends the information of a previous publication by the IAEA (Molecular Diagnostics PCR Handbook, Viljoen, G.J., Nel, L.H., and Crowther, J.R., Springer, 2005) and deals with the basic principles of both PCR and real-time PCR with a view to their use for the early and rapid diagnosis of harmful pathogens of animals and those of zoonotic nature origin to form the basis of real-time PCR technologies and applications.

E.A. Pestana et al., *Early, Rapid and Sensitive Veterinary Molecular Diagnostics - Real Time PCR Applications*, DOI 10.1007/978-90-481-3132-7_1,

The transfer of PCR technology is challenging and often made without adequate consideration of the way it is set up; the training of staff and as to exactly how it is to be used. Implementation of this modern technology, that promises so much, holds many dangers and conventional techniques should never be ignored at the expense of the PCR alone. A well-considered argument as to why PCR should be set up, along with considerations of the cost benefit in the short and long term, is necessary. This book is intended to give a concise overview of the practical aspects of PCR considering its best use in terms of laboratory practice; the setting up of laboratories to perform PCR; and GLP and standardisation of PCR protocols. Difficulties, as well as advantages in PCR application, are highlighted dealing with more recent advances in methods and putting the PCR in context of other tests. It is hoped that the book will appeal to readers of all disciplines and levels so that they can better judge the merits of the techniques and develop the technology in a more logical and efficient way. The book is detailed in many areas as a direct help to those involved in everyday PCR. Specific examples are given of protocols used as standard operating procedures (SOP's) in individual laboratories; routine uses; latest developments and potential of PCR technologies and training material necessary for PCR technology transfer. Specific emphasis is placed on the use of PCR for diagnosis of infectious diseases, aimed at diagnosticians and scientists of the developing world. All patents and licences, directly or indirectly related to the technologies and processes addressed in this book should be respected.

1.2 What Is PCR?

Molecular biology has been revolutionised by PCR, a method that efficiently increases the number of DNA molecules in a logarithmic and controlled fashion. The concept of DNA amplification by PCR is simple and its impact has been extraordinary. Kary Mullis conceived PCR in 1983 and the first PCR publication appeared in 1985. Every year thereafter, the number of papers on PCR has risen exponentially

In 1989, Science magazine selected PCR as the "major scientific development" and Taq polymerase, the enzyme essential to PCR's success, as "molecule of the year". In 1993, Kary Mullis received the Nobel prize for chemistry. The advent of PCR meant that insufficiencies in the quantity of DNA were no longer a limitation in molecular biology research or diagnostic procedures. It is indeed difficult to find publications in the biological sciences that do not describe the application of PCR in some or other way. The chemistry involved in PCR depends on the complementarities (matching) of the nucleotide bases in the double-stranded DNA helix. When a molecule of DNA is sufficiently heated, the hydrogen bonds holding together the double helix are disrupted and the molecule separates or denatures into single strands. If the DNA solution is allowed to cool, the complementary base pairs can reform to restore the original double helix. In order to use PCR, the exact sequence of nucleotides that flank (lay on either side of) the area of interest (the target area that needs to be amplified), must be known. This is the absolute minimum data necessary before a typical PCR reaction can be used. This data is necessary for the design of PCR primers that are 5′-3′ oligonucleotides of about 20 nucleotides in length. These are designed to be complementary to the flanking sequences of the target area, as mentioned previously. Thus, the researcher has to either use previous data (known information of sequences) or, if this is unavailable, determine the sequence of these regions experimentally. The two primers (primer pair) can then be synthesized chemically and will then serve as leaders or initiators of the replication step. The key to the replication reaction is that it is driven by a heat-stable polymerase molecule that reads a template DNA in the 3′-5′ direction and synthesises a new complementary template in the 5′-3′ direction, using free dideoxy nucleoside triphosphates (dNTP's = nucleotide bases) as building blocks.

1.3 What Is the Use of PCR?

PCR is primarily a method to spectacularly amplify a desired DNA fragment (piece of DNA) in order to increase the target DNA to detectable levels. This has had a profound effect on all molecular studies including those in the diagnostic area. It suddenly changed the way sensitivity was defined, as we are now able to detect very low numbers of pathogens with great accuracy. We are also able to detect carrier animals more easily, detect mixed populations of pathogens in an infection and to determine pathogen load. The method has found numerous related applications in molecular biology and now forms the fundamental basis of most studies involving genetic material.

As illustration of this uniqueness, PCR can be used very effectively to modify DNA. Such modification may include the addition of restriction enzyme sites (in order to facilitate cloning requirements) or regulatory elements (e.g., the addition of promoter sequences to a DNA cistron). A further type of modification can be the generation of desired site-directed mutations in a gene, inclusive of sequence alterations, additions or deletions. Cycle-sequencing, a modification of the classical di-deoxy sequencing method pioneered by Fred Sanger in the early 1980s, uses the principles of PCR to rapidly perform sequence reactions in a thermal cycler. Equally often used is the sensitive protein-DNA or protein-RNA interaction analysis (e.g., for the activation of Polymerase type II promoters) approach of the Maxim and Gilbert sequence reactions, using nuclear applications to study protein and nucleic acid interactions on a molecule by molecule basis. The way in that PCR has dramatically impacted on diagnosis of genetic and infectious disease is one of the foci of this book. For PCR-directed diagnostics it is possible to work with crude samples and minute amounts of material that may include degraded templates, blood, sperm, tissue, individual hairs, etc. In related applications, PCR plays a central role in genetic typing and molecular characterization of organisms or individuals and molecular epidemiology. Some examples of the general application of PCR in molecular biology are given in **Table 1.1** and specific focus is given to the diagnostic uses of this technique in further sections of the book.

Table 1.1 General applications of PCR

1. Diagnosis of pathogens	(a) PCR (b) Nested PCR (c) Quantitative PCR (d) Multiplex PCR (e) Differential on-line and real time PCR
2. Typing genetic markers	(a) RFLPs (b) AFLPs (c) Short tandem repeat polymorphisms (d) SNPs
3. DNA template for	(a) Genomic mutation screening intron-specific primers flanking exons (b) RT-PCR cDNA used as templates for pairs of exon-specific primers to generate overlapping fragments
4. Detecting point mutations	(a) Restriction site polymorphisms (b) Allele specific amplification
5. cDNA cloning	(a) DOP-PCR (b) RACE
6. Genomic DNA cloning	New members of a DNA family (a) DOP-PCR. Whole genome or subgenomic amplification (b) DOP-PCR (c) Linker-primed PCR
7. Genome walking	(a) Inverse PCR (b) Bubble linker (vectored) PCR (c) IRE-PCR
8. DNA templates for DNA sequencing	(a) ssDNA by asymmetric PCR (b) dsDNA for direct sequencing or for cloning followed by sequencing
9. In vitro mutagenesis	(a) 5' add-on mutagenesis to create a recombinant PCR product (b) Mispaired primers to change a single predetermined nucleotide

1.4 PCR and Infectious Diseases – The Veterinary Picture

Infectious diseases can be caused by microbial pathogens, include agents of fungal, protozoan, bacterial, clamydial, rickettsia and viral nature. Despite many advances in diagnostics and vaccinology, infectious diseases still have devastating consequences for agricultural, economies, worldwide.

Three examples of devastation with regard to animal husbandry since the 1990s, include the emergence of the prion, bovine spongiform encephalopathy (BSE); the huge outbreaks of foot-and-mouth disease (FMD) in Europe and avian influenza (AI) in Asia and elsewhere. A great many of these infectious diseases can be transmitted from vertebrate animal to man (called zoonoses) where more than 200 such zoonotic diseases are known. Infectious diseases are typically transmitted through the skin or eyes (direct contact, insect vectors, bite wounds, sexual contact). In other cases agents are airborne and infect the epithelial cells lining the respiratory tract from where further systemic infection may proceed. Additional sources of infectious microorganisms are contaminated food and water with a route of infection through the mouth and alimentary tract, or through the respiratory system.

Infectious diseases can be considered new, emerging, re-emerging or resurgent. These are diseases that have been described in the last 10–30 years, or that are caused by specific modifications of agents already present in the environment; e.g., in a different host reservoir). These agents evolve, mutate or are otherwise epidemiologically affected by changing conditions or other selective advantage.

Typically, re-emerging diseases are those that have persisted at a subdued level in the population and recur as a result of antimicrobial drug resistance or other changes that might favor dramatic increases in disease incidence. Re-emerging diseases can also be described as resurgent, referring to an abrupt increase in incidence or geographic distribution of the particular disease. The emergence and re-emergence of diseases are clearly related to changes in the infectious pathogen, the vector or transmission system and the host population.

Apart from drug resistance, other epidemiologically important changes include mutations that lead to increased virulence; changes in the distribution or activity of vectors; globalization and increased travel; war, population explosions; climatic and ecological changes; geographical displacement of species; movement into previously uninhabited areas; poverty and breakdown of animal or healthcare systems; and changes in agriculture and industrialisation.

The quest for improved diagnostic methods to combat infectious diseases has become ever more demanding. As our knowledge of the structure and function of pathogens and of the immunological responses of a host to infection or antigenic stimulation expands; approaches to both disease diagnosis and to immunisation has changed from empirical methodologies to procedures undertaken in a much more informed manner. These approaches have enabled diagnostic assays to become more specific and better standardized and in the development of safer and more effective quality controlled vaccines. Advances in laboratory technology have enabled diagnostic assays to be performed and analysed with a greater degree of automation, with improved precision and reduced labour requirements. However, some of these advances have come at an increased cost and it is therefore not always appropriate to apply them, especially with the strong budgetary constraints of most developing countries. We now have the scientific knowledge and technical skills available to make very significant further advances in the diagnosis of, and immunisation against, specific diseases. Such expected advances will also be highlighted in the following discussion concerning laboratory diagnosis and technology.

1.5 Laboratory Diagnostic Technology

Laboratory diagnostic technology is directed towards either,

The detection of the presence, or absence, of a pathogen and its subsequent identification and characterization

The detection of the pathological effect of, or immunological response to, infection by a particular pathogen

In the past, the detection of pathogens was achieved by visualisation of the organism by light or electron microscopy, either directly in specimens from the affected animal or following culturing.

Alternatively, serological procedures are used where a specific antigen is detected using characterized antibodies and significant advances have been made in such serological procedures. There have also been enormous advances in our understanding of immunology. With a few exceptions, immunological approaches to infectious disease diagnosis are directed towards serological procedures, i.e., the detection of interactions between antigens unique to the pathogen and specific antibodies elaborated against them. Grouping or typing of pathogens through serology has played a major role in pathogen classification and in epidemiological studies.

More recently with the advances in molecular biology, the analysis of the genetic material of pathogens has complemented or even replaced serological methods for diagnostics, epidemiology and taxonomy. Fragments of DNA or RNA, produced by nuclease digestion, can be separated by electrophoresis to form bands whose position in a gel is dictated by their molecular size. The patterns obtained for digests from different sources can be compared by running them in parallel. Greater discrimination can be obtained by running gels in two dimensions, under conditions producing separation in the first dimension on the basis of charge and in the second on the basis of size. Characteristic patterns are obtained (fingerprints) that can be compared visually. This can be used for comparing isolates of a particular pathogen.

There are significant advantages in the ability to indicate a pathogen's presence by the detection of its DNA or RNA. Successful bacterial or viral isolation is dependent on the presence of live or viable pathogen in a specimen and is generally time consuming and expensive. It also requires the presence of live pathogen. Antigen detection procedures are limited by the amount and quality of antigen present in specimen. Nucleic acid is more resistant to denaturation than protein and can survive long period of time (even centuries) under appropriate conditions. The limitation on nucleic acid detection has been due to the very small amounts available for detection. Notwithstanding, nucleic acid hybridisation techniques have been used to probe specimens, using complementary strands of DNA or RNA appropriately labelled with an enzyme or a radioisotope. Specific base pairing produces a hybrid between the probe and the target that can be detected through the specific label. Such nucleic acid probes have been developed and used for the detection of many pathogens.

PCR represents an entirely new technology. In vitro bacterial or viral culture is widely used to isolate and multiply pathogens, so that the organism itself, or its antigens, can be more readily detected, by being present in greater quantity and generally with fewer contaminants. PCR technology permits the same principle (i.e., in vitro amplification) to be applied to the detection of specific sequences of nucleic acid. There are enormous benefits to this approach.

The application of PCR to disease diagnosis has been somewhat restricted to laboratories with the required facilities, equipment, funding and expertise. The procedure must be made in very clean conditions since contamination with minute amounts of extraneous DNA may produce false positive results. Often this means that separate rooms and/or laminar flow cabinets must be used with careful disinfection protocols to avoid cross-contamination. Equipment costs are high but become justifiable to more laboratories as the range of applications increase.

Methods are becoming better standardised so that training of staff in appropriate techniques can be rationalised. Certain standard reagents and consumables, such as Taq DNA polymerase, are expensive and not readily available in some countries. Nevertheless, it can be expected that these limitations will be progressively (probably rapidly) addressed and that this technology will be applied to an increasing extent.

The application of PCR and other hybridisation techniques to diagnostics is dependent on gene sequencing. Sequences that are unique to the pathogen must be targeted and the regions chosen for complementary primer production must be conserved within the genotypic range of the pathogen. Thus the challenge for many laboratories is have access to specific primers and oligonucleotide detection probes. While these can be synthesised in suitably equipped laboratories, for most laboratories in developing countries there will be a dependence on having these reagents custom made by commercial sources or accessing them from international or national reference laboratories.

A large number of specific diagnostic PCR methods have been made available in the public domain though publications in scientific journals, websites and the like. It is beyond the scope of this book to provide a comprehensive list of such methods. However, a few diagnostic assays for important veterinary diseases for which established standard operating procedures (SOPs) have been described and which are widely implemented, have been included in Chapter 5. These include detailed methods for Avian Influenza, Swine vesicular disease, African swine fever, Rift valley fever (RVF), African horse sickness (AHSV), Bluetongue (BTV), CSFV, and foot and mouth disease (FMD).

Bibliography

1. Armstrong, RM, Samuel, AR, Carpenter, WC, Kant, R, Knowles, NJ. 1994. A comparative study of serological and biochemical methods for strain differentiation of foot-and-mouth disease type A viruses. *Vet. Microbiol.*, **39**, 285–98
2. Chamberlain, RW, Wamwayi, HM, Hockley, E, Shaila, MS, Goatley, L, Knowles, NJ, Barrett, T. 1993. Evidence for different lineages of rinderpest virus reflecting their geographic isolation. *J. Gen. Virol.*, **74**, 2775–80

3. Dedieu, L, Mady, V, LeFevre, PC. 1994. Development of a selective polymerase chain reaction assay for the detection of Mycoplasma mycoides subsp. mycoides S.C. (contagious bovine pleuropneumonia agent). *Vet. Microbiol.*, **42**, 327–39
4. Donn, A., Martin, LA, Donaldson, AI. 1994. Improved detection of persistent foot-and-mouth disease infection in cattle by the polymerase chain reaction. *J. Virol. Meth.*, **49**, 179–86
5. Dunn, CS, Donaldson, AI. 1997. Natural adaptation to pigs of a Taiwanese isolate of foot-and-mouth disease virus. *Vet. Rec.*, **141**, 174–5
6. Forsyth, MA, Barrett, T. 1995. Evaluation of a polymerase chain reaction for the detection and characterisation of rinderpest and peste des petits ruminants for epidemiological studies. *Virus Res.*, **39**, 151–63
7. Harding, MJ, Prud'homme, I, Gradil, CM, Heckert, RA, Riva, J, McLaurin, R, Dulac, GC, Vydelingum, S. 1996. Evaluation of nucleic acid amplification methods for the detection of hog cholera virus. *J. Vet. Diagn. Invest.*, **8**, 414–9
8. Ireland, DC, Binepal, YS. 1998. Improved detection of capripoxvirus in biopsy samples by PCR. *J. Virol. Meth.*, **74**, 1–7
9. King, AMQ, Underwood, BO, McCahon, D, Newman, JWI, Brown, F. 1981. Biochemical identification of viruses causing the 1981 outbreaks of foot and mouth disease in the UK. *Nature*, **293**, 479–80
10. Meyer, RF, Brown, CC, House, C, House, JA, Molitor, TW. 1991. Rapid and sensitive detection of foot and-mouth disease virus in tissues by enzymatic RNA amplification of the polymerase gene. *J. Virol. Meth.*, **34**, 161–72
11. Palmenberg, AC. 1989. Sequence alignments of picornaviral capsid proteins. In: Molecular Aspects of Picornaviral Infection and Detection. Eds. Semler, BL. and Ehrenfeld, E. American Society for Microbiology, Washington, DC. pp. 211–41
12. Saiki, RK, Gelfand, DH, Stoffel, S, Scharf, SJ, Higuchi, R, Horn, GT, Mullis, KB, Erlich, HA. 1998. Primer-directed enzymatic amplification of DNA with a thermostable DNA polymerase. *Science*, **239**, 487–91

Chapter 2
Traditional PCR

Contents

2.1 Traditional PCR

Polymerase chain Reaction (PCR) was a method invented by Kary Mullis, who was awarded the Nobel Prize in Chemistry in 1993 for his innovation. In the 1980s, Mullis, who was working at Cetus Corporation, imagined a way to start and stop a polymerase's action at specific points along a single strand of DNA. Cetus scientists eventually succeeded in artificially multiply DNA through repeated cycles of duplication driven by an enzyme called DNA polymerase. A polymerase is a naturally occurring enzyme, a biological macromolecule that catalyzes the formation and repair of DNA (and RNA); activity scientists have learned to manipulate.

E.A. Pestana et al., *Early, Rapid and Sensitive Veterinary Molecular Diagnostics - Real Time PCR Applications*, DOI 10.1007/978-90-481-3132-7_2,

Science Magazine named PCR and the polymerase "Molecule of the Year" in 1989, the editor, Daniel Koshland Jr., defined what PCR is as follows, "The starting material for PCR, the 'target sequence,' is a gene or segment of DNA." In a matter of hours, this target sequence can be amplified a million fold. The complementary strands of a double-stranded molecule of DNA are separated by heating. Two small pieces of synthetic DNA, each complementing a specific sequence at one end of the target sequence, serve as primers. Each primer binds to its complementary sequence. Polymerases start at each primer and copy the sequence of that strand. Within a short time, exact replicas of the target sequence have been produced. In subsequent cycles, double-stranded molecules of both the original DNA and the copies are separated; primers bind again to complementary sequences and the polymerase replicates them. At the end of many cycles, the pool is greatly enriched in the small pieces of DNA that have the target sequences, and this amplified genetic information is then available for further analysis.

2.2 PCR Reaction

When performing a PCR reaction, several components are needed. The first one is the DNA template, which contains the region of the DNA fragment to be amplified. To achieve this, two primers, which determine the beginning and end of the region to be amplified by the DNA polymerase, are designed. The DNA polymerase, then will built a new DNA strand, and to that may need of other components such as PCR Buffer, which provides a suitable chemical environment for the polymerase to perform the amplification; dNTPs (deoxyiribosenucleotides), which are the monomers that DNA polymerase uses to form DNA; and a co-factor to increase the yield of the reaction (magnesium Chloride). Each of these PCR components will be discussed later on detail.

Basic elements of reaction mixture for PCR	
Template DNA	10^5–10^6 target molecules
Primer	Between 0.1 and 0.5 μM each
10x Buffer	As supply with the enzyme
$MgCl_2$	1.5 mM
dNTPs	200 μM of each dATP, dCTP, dGTP and dTTP
Heat-stable DNA polymerase	1–2 Units

The PCR process consists of a series of 20–30 cycles. Each cycle consists of three steps:

(1) The double-stranded DNA has to be heated to 94–96°C in order to separate the strands. This step is called melting; it breaks apart the hydrogen bonds that connect the two DNA strands. Prior to the first cycle, the DNA is often melted for an extended time to ensure that both the template DNA and the primers have completely separated (single-strand).

(2) After separating the DNA strands, the temperature is lowered so the primers can attach themselves to the single DNA strands. This step is called annealing. The temperature of this stage depends on the primers and is usually 5°C below their melting temperature (45–60°C). A wrong temperature during the annealing step can result in primers not binding to the template DNA at all, or binding at random.

(3) Finally, the DNA-Polymerase has to fill in the missing strands. It starts at the annealed primer and works its way along the DNA strand. This step is called elongation. The elongation temperature depends on the DNA polymerase. The time for this step depends both on the DNA polymerase itself and on the length of the DNA fragment to be amplified; which is usually 1 min per 1000 bp. After the PCR reaction is complete, PCR products can be identified by its size using agarose gel electrophoresis. As a result, the smaller DNA strands move faster than the larger strands through the gel toward the positive current. The size of the PCR product can be determined by comparing it with a DNA ladder, which contains DNA fragments of known size, also within the gel.

2.2.1 Primer Specifications

PCR requires a DNA template and a pair of primers flanking the target DNA. An important parameter to be considered when selecting PCR primers is the ability of the primers to form a stable duplex exclusively with the specific site to be targeted on the DNA molecule. The melting temperature of the primers, which is defined as the temperature below which the primer will anneal to the DNA template and above which the primer will dissociate (break apart) from the DNA template; usually depends on the length of the primer designed. The length of the primers should be usually between 18 and 24 bp. If the primers are designed too short, the probability of them annealing at different region on the DNA template usually increases; whereas if primers are too long, their melting temperature would also increase, and thus having a tendency for secondary annealing. Primers with melting temperatures in the range of 52–58°C generally produce the best results. The concentration of each primer should be between 0.1 and 0.5 μM. For most applications 0.2 μM produces satisfactory results. Too high primer concentrations increase the chance of mispriming, which results in nonspecific PCR products. Limiting primer concentrations result in extremely inefficient PCR reactions.

Primer design guidelines
• *GC-content* (the number of G's and C's in the primer as a percentage of the total bases) should be between 40–60%.
• Tm for both primers used in the reaction should not differ >5°C and the Tm of the amplification product should not differ from primers by >10°C.
• *GC Clamp formation.* More than 3 G's or C's should be avoided in the last 5 bases at the 3′ end of the primer.
• *Secondary Structures. Hairpins* (intramolecular interaction within the primer), *self-dimers* (intermolecular interactions between the two (same sense) primers), *and cross-dimers* (intermolecular interaction between sense and antisense primers) *should be avoided.* Inner self-complementary hairpins of >4 and of dimers >8 should be avoided.
• *Repeats* should be avoided because they produce mispriming. A maximum number of di-nucleotide repeats *acceptable are 4 di-nucleotides.*
• *3′ terminus* is should be carefully considered – it must *not be complementary* to *any region* of the other primer used in the reaction and must provide correct base matching to template.

2.2.2 DNA Template

The DNA template used for PCR is often overlooked when compared with the effort put into primer design.

The most commonly used parameters that relate to the DNA template are the PCR product size and the Tm of the product. However, it is known that DNA templates with a very high or very low GC/AT ratio can be difficult to amplify.

Usually the amount of template DNA used on a PCR reaction is in the range of 0.01–1 ng for plasmid or phage DNA and 0.1–1 μg for genomic DNA, for a total reaction mixture of 50 μL. Higher amounts of template DNA usually increase the yield of non-specific PCR products.

Too much template may also decrease efficiency due to contaminants in the DNA preparation. These include urea, the detergent SDS (whose inhibitory action can be reversed by nonionic detergents), sodium acetate, and, sometimes, components carried over in purifying DNA from agarose gels.

Inhibition of PCR by impurities on the template

Substance	Inhibitory concentration
SDS	>0.005% (w/v)
Phenol	>0.2% (v/v)
Ethanol	>1% (v/v)
Isopropanol	>1% (v/v)
Sodium acetate	>5 mM
Sodium chloride	>25 mM
EDTA	>0.5 mM

2.2.3 dNTPs

The concentration of each dNTP (dATP, dCTP, dGTP and dTTP) in the reaction mixture is usually 200 μM. Too high concentrations of dNTPs usually inhibit PCR reactions.

It is very important to have equal concentrations of each dNTP (dATP, dCTP, dGTP, dTTP), as inaccuracy in the concentration of even a single dNTP dramatically increases the misincorporation errors.

2.2.4 Magnesium Chloride

The concentration of Magnesium chloride can have a particular effect on the specificity and yield of the PCR reaction. Its concentration must be optimized for every primer/template system. Usually few Mg^{2+} ions result in a low yield of PCR product, on the other hand too many increase the yield of non-specific products and promote misincorporation errors (see figure below). When the DNA samples contain EDTA or other chelators, the $MgCl_2$ concentration in the reaction mixture should be raised proportionally. Many components of the reaction bind magnesium ion, including primers, template, PCR products and dNTPs. The main 1:1 binding agent for magnesium ion is the high concentration of dNTPs in the reaction. Because it is necessary for free magnesium ion to serve as an enzyme cofactor in PCR, the total magnesium ion concentration must exceed the total dNTP concentration. For example if in the optimization process we start with a concentration of 1.5 mM magnesium chloride in the presence of 0.8 mM total dNTPs; it would mean that there is still about 0.7 mM free magnesium left to be bind to the DNA polymerase. In general, magnesium ion should be titrated in a concentration series from 1.5 to 4.0 mM in 0.5 mM steps.

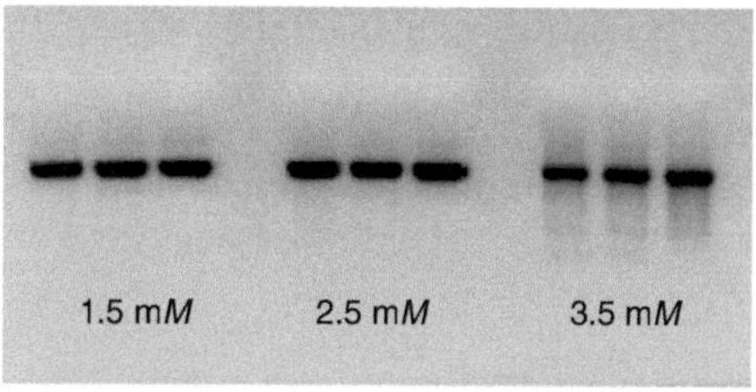

2.2.5 DNA Polymerase

The PCR reaction conditions and reaction times depend on the type of DNA polymerase used. Usually, when using *Taq* DNA polymerase, the recommended concentration of polymerase is 1–1.5 u in 50 μL of reaction. Higher *Taq* DNA polymerase concentrations may cause synthesis of non-specific products. However, if inhibitors are present in the reaction mix (e.g., if the template DNA used is not highly purified), higher amounts of *Taq* DNA polymerase (2–3 u) may be necessary to obtain a better yield of amplification products.

2.2.6 Polymerase Buffer

All DNA polymerases are supplied with their own optimal polymerase buffer. The standard buffer usually includes 10–50 mM Tris-HCl (pH 8.3–8.8), and up to 50 mM KCl may be included to facilitate primer annealing.

Some suppliers of DNA polymerases have added NH_4^+ ions to their buffers. It has been shown that the presence of NH_4^+ ions results in a high specificity of the primer-template binding over a broad temperature range. Others also contain NaCl, which is usually used for GC rich templates (>60%).

Although, little is known of the exact role of these chemicals in PCR, good results have been also obtained by the addition of glycerol, DMSO (5–20%), formamide (5–20%) or tetramethylammonium chloride (0.01–10 mM) to the reaction mix.

2.2.7 Cycling Conditions

Amplification parameters in each cycle depend strongly on the design of the thermocycler and primers and template.

2.2.7.1 Initial Denaturation Step

The complete denaturation of the DNA template at the start of the PCR reaction is of key importance. Incomplete denaturation of DNA results in the inefficient utilization of template in the first amplification cycle and in a poor yield of PCR product. The initial denaturation should be performed over an interval of 1–3 min at 95°C if the GC content is 50% or less. This interval should be extended up to 10 min for GC-rich templates.

2.2.7.2 Denaturation Step

Usually denaturation for 0.5–2 min at 94–95°C is sufficient, since the PCR product synthesized in the first amplification cycle is significantly shorter than the template DNA and is completely denatured under these conditions. If the amplified DNA has a very high GC content, denaturation time may be increased up to 3–4 min. Alternatively, additives facilitating DNA denaturation – glycerol (up to 10–15 vol.%), DMSO (up to 10%) or formamide (up to 5%) should be used. In the presence of such additives, the annealing temperature should be adjusted experimentally, since the melting temperature of the primer-template DNA duplex decreases significantly when these additives are used. The amount of enzyme in the reaction mix should be increased since DMSO and formamide, at the suggested concentrations, inhibit *Taq* DNA Polymerase by approx. 50%. Alternatively, a common way to decrease the melting temperature of the PCR product is to substitute dGTP with 7-deaza-dGTP in the reaction mix.

2.2.7.3 Primer Annealing Step

Usually the optimal annealing temperature is 5°C lower than the melting temperature of primer-template DNA duplex. Incubation for 0.5–2 min is usually sufficient. However, if non-specific PCR products are obtained in addition to the expected product, the annealing temperature should be optimized by increasing it stepwise by 1–2°C. As seen in the picture, the effect of the annealing temperature is critical on the success of the PCR reaction. Several annealing temperatures for the same primer pair, were tested. Lower annealing temperatures give raise to unspecific and poor PCR products, whereas using the correct annealing temperature (line 8) a clear PCR product could be resolved.

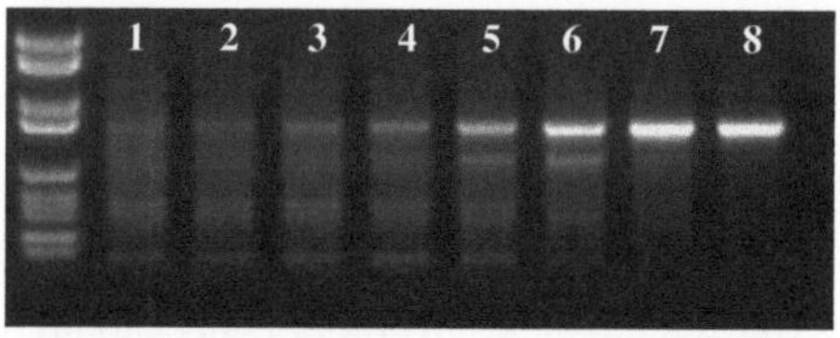

2.2.7.4 Extension Step

Usually the extension step is performed at 72°C. The rate of DNA synthesis by *Taq* DNA polymerase is highest at this temperature. Recommended extending time is 1 min for every Kb to be synthesised. When larger DNA fragments are amplified, the extending time is usually increased by 1 min for each 1000 bp.

2.2.7.5 PCR Amplification Cycle Number

Cycle number vs. starting material concentration

Number of target molecules	**Number of cycles**
3×10^5	25–30
1.5×10^4	30–35
1×10^3	35–40
50	40–45

The number of cycles necessary to obtain a sufficient amount of PCR product depends strongly on the concentration of the DNA template. In a typical PCR, the maximum amount of product is approx. 10^{12} copies of the template. Starting from one copy, the most efficient PCR would reach this level in 40 cycles. Depending on the nature of the DNA template, as a rule of thumb 25 cycles should be sufficient for plasmid DNA and 30–35 cycles for genomic DNA. Too many cycles can increase the amount and complexity of non-specific background products (see Plateau Effect). Of course, too few cycles will produce low PCR product yield.

2.2.7.6 Final Extension Step

After the last amplification cycle, samples are usually incubated at 72°C for 5–15 min to fill-in the protruding ends of newly synthesized PCR products. Also, during this step, the terminal transferase activity of the Taq DNA polymerase adds extra A nucleotides to the 3′-ends of PCR products. This step is important when PCR fragments are to be cloned into T/A vectors; if so, this step can be prolonged to up to 30 min.

2.3 PCR Set Up and Optimization

The master mix typically contains all of the components needed for PCR except the template DNA. Before starting the PCR, all components should be gently mixed and centrifuge before proceeding to the mixing. The table below shows a typical PCR reaction composition for a final volume of 50 μL.

PCR component	50 μL master mix	Final concentration
Sterile deionized water	Variable	
10X *Taq* buffer	5 μL	1X
2 mM dNTP mix	5 μL	200 μM of each dNTP
Primer I	Variable	0.1–0.5 μM
Primer II	Variable	0.1–0.5 μM
Taq DNA Polymerase	Variable	1–1.5 u
25 mM $MgCl_2$	Variable	1–4 mM
Template DNA	Variable	10 pg–1 μg

Should Magnesium chloride be added to the reaction, the following table provides a selection of concentration to be used during titration, using a stock solution of 25 mM $MgCl_2$.

Final concentration of $MgCl_2$ in 50 μL reaction mix (mM)	1.0	1.25	1.5	1.75	2.0	2.5	3.0	4.0
Volume of 25 mM $MgCl_2$ (μL)	2	2.5	3	3.5	4	5	6	8

2.3.1 Optimizing a PCR Reaction

Some companies offer a PCR optimization kit that may be use to simplify the PCR optimization procedure. In addition, approaches such as the touchdown PCR also offers simple one-step optimization of PCR reactions that are expected to be sub-optimal with regard to primer/template homology.

As a general rule, however, any PCR that will become an established assay in the laboratory should be properly optimized by a titration method. As described above, PCR conditions are unique for each assay and are influence by all the components present on the reaction mix.

Although, the key to successful PCRs lies in the design of appropriate primers, the parameters described above for each PCR component (dNTPs, Mg^{+2}, and enzyme) may also play an important role on the optimization of the PCR.

2.4 The PCR Plateau Effect

Theoretically, the amount of PCR product doubles during each cycle of the PCR reaction, but, in reality, beyond a certain number of cycles, the efficiency of amplification decreases with increasing cycle number, resulting in the plateau effect, as shown in Fig. 2.1.

A number of factors may cause the plateau effect, including,

1. Degradation of nucleotides or primers.
2. Inactivation of the DNA polymerase enzyme (half-life is 40 min at 95°C).
3. Re-association of single stranded PCR fragments before primers can anneal or be extended.
4. Substrate excess where there is more DNA than the amount of enzyme available to replicate it in the allotted polymerization time.
5. Competition by non-specific amplification products.
6. Accumulation of inhibitors of polymerase activity, such as pyrophosphates.

The number of PCR cycles at which the plateau effect occurs varies greatly with the particular DNA sequence being amplified. Length, GC content, and the presence of any secondary structure in the sequence to be amplified are all important, as is the initial total quantity and concentration of the target DNA. As a result, the number of cycles at which the plateau effect occurs must be individually and empirically determined for each target sequence.

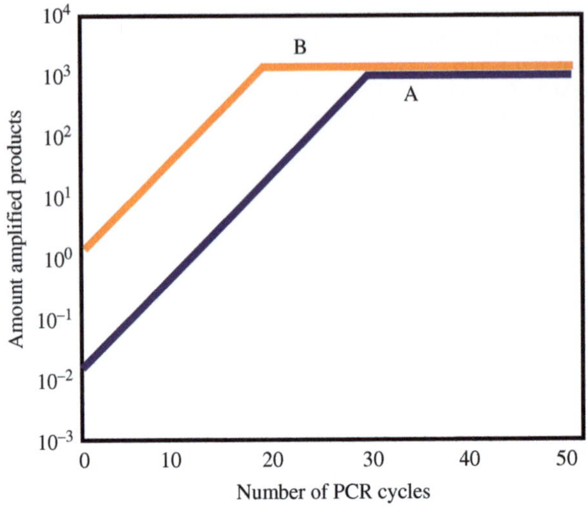

Fig. 2.1 The plateau effect in PCR

The end-point or Plateau phase should not be used to compare differences in target quantities, because differences observed in the amount of detectable product after the plateau effect has been reached are frequently artifacts and may show little relationship to the quantity of starting target material. Therefore, end point quantification of PCR products in plateau phase is unreliable. Only during the exponential phase of the PCR reaction is it possible to extrapolate back to determine the starting quantity of the target sequence contained in the sample. This is the attribute of PCR that makes Real-Time Quantitative PCR so necessary. This important difference between Traditional vs. Quantitative Real Time PCR will be discussed in the next chapter.

Finally it is important to quote that an important consequence of reaching plateau is that an initially low concentration of non-specific products resulting from mispriming events may continue to amplify preferentially. Therefore, optimizing the number of PCR cycles is the best way to avoid amplifying background products.

2.5 Radioisotope-PCR Based Methods

PCR together with other molecular biology techniques is being extensively used on the identification of molecular markers for disease detection or more specifically pathogen detection and diagnosis; development of diagnostic tests, gene expression (i.e. cytokines, growth factors, transcription factors, etc.), allelic discrimination (detection of single nucleotide polymorphism, SNP) and genotyping.

Ever since the invention of molecular biology techniques, radionuclide based methods have been an integral component of their development and they are important tools for their application in diagnosis and research. Radionuclide based molecular techniques have proved to be highly sensitive, specific, robust and cost effective in various application contexts.

Radioisotopes are used particularly during the detection step, when a need for high sensitivity and specificity on a test are required. This is due to the fact that radioisotopes allow detection of minimal quantities of DNA (0.1 pg). They are also needed when specificity and sensitivity are required during hybridization, a widely use technique in molecular biology. Moreover, radioisotopes provide a high sensitivity (125-fold) compared for example with the popular staining systems in molecular biology, such as ethidium bromide. Also, compared to fluorescent or enzymatic labels, radioisotopes provide a good "signal to noise" ratio and higher specificity. The strong signal emitted by radiolabels stands out against the background as compared to the signal emitted from a fluorescent probe. The innate fluorescence in many biological samples may also lead to high backgrounds in hybridization experiments, e.g., due to the presence of biotin in tissues, when using for example a biotinylated probe with streptavidin, a problem of poor signal-to-background ratio usually occurs.

Several PCR-based molecular biology techniques e.g., dot blot assay, restriction fragment length polymorphism (RFLP), single stranded conformational polymorphism (SSCP), amplified fragment length polymorphism (AFLP), differential-display reverse transcription-PCR, mismatch cleavage assay, heteroduplex tracking assay (HTA), DNA sequencing, microsatellite detection, macroarray chip technology, make use of isotopes.

Many radioisotope-based methods can be described as important tools for animal diseases research and diagnosis. Despite the emergence of alternative methods, as explain above radioisotopes still offer a better chance when sensitive and specific assays are required.

2.5.1 Radioisotopic-Based Methods

Various molecular biology methods that use isotopes are described in this section. They are based on PCR amplification in order to work with a measurable amount of DNA.

2.5.1.1 PCR Dot-Blot Assay

Dot-blot is a method quite similar to Southern or Western Blot, but simplified. As the technique does not require an electrophoresis step, samples are transferred or blotted after PCR onto a single membrane as separate dots. The membrane is then treated with an appropriate radioactive single stranded DNA probe under conditions favouring the hybridization. Dots having the appropriate DNA or RNA sequence will hybridize with the radioactive probes. These dots are detected by autoradiography; the intensity of dot in the autoradiograph corresponds fairly well with the extent to which DNA or RNA is represented in the sample. Dot-blot assays has been used for determining the sex of a variety of species of birds, diagnostics of types HIV-1 and HIV-2, and SNP discrimination.

2.5.1.2 DNA Sequencing

In 1975, Sanger introduced his "plus and minus" method for DNA sequencing. This was a critical transition technique leading to the modern generation of methods that have completely dominated sequencing over the past 30 years.

The key to this advance was the use of polyacrylamide gels to separate the isotopic labelled products of primed synthesis by DNA polymerase in order of increasing chain length. The method analyzed the products of DNA polymerase reactions that extended a primer annealed to a single-stranded DNA template. DNA synthesis to extend the primer was carried out in two sequential DNA polymerase reactions. The first was carried out under conditions where synthesis was slow and asynchronous, resulting in a population of all possible products extending 1,2,3,... up to a few hundred bases. A ^{32}P labelled nucleotide was incorporated at this step. This product was then divided into eight aliquots and used to prime a second round of DNA polymerase reactions. In these reactions, synthesis was terminated in a sequence-specific manner by supplying only one of the four nucleoside triphosphates ("plus" reactions), or else three of the four ("minus" reactions). The eight reactions were electrophoresed in adjacent lanes of a 12% acrylamide, 8 M urea denaturing gel. Following electrophoresis the gel was placed in contact with X-ray film for a suitable time, typically overnight. When the film was developed, molecules differing by a single nucleotide in length could be resolved as discrete bands on the resulting autoradiograph. This allowed a sequence of 50 bases to be deduced in a single experiment. The main problem with the method is the difficulty in determining the length of homopolymer runs. Bands corresponding to the beginning and end of such runs are produced, but no bands are produced for positions internal to runs, so run lengths must be estimated from band spacing in the gel. This becomes unreliable for longer runs.

Maxam and Gilbert developed a DNA sequencing method that was similar to the Sanger and Coulson method in using polyacrylamide gels to resolve bands that terminated at each base throughout the target sequence, but very different in the way that products ending in a specific base were generated. Their method started with a double-stranded DNA restriction fragment radiolabeled at one end with ^{32}P. The fragment was then cleaved by base-specific chemical reactions. One reaction cleaves at both purines (the "A + G" reaction), one preferentially at A ("A > G"), one at pyrimidines ("C + T") and one at cytosines only ("C"). Unlike the plus and minus method, the chemical method produced bands for every sequence position, including those within homopolymer runs. This advantage led to early widespread adoption of the chemical method following its publication in February 1977.

The problems with the plus and minus method were solved when Sanger developed "the dideoxy method" and published it in December 1977. The underlying concept was to use chain-terminating nucleotide analogs rather than subsets of the four natural dNTPs to cause base-specific termination of primed DNA synthesis. In the original implementation both arabinoside triphosphates and 2′,3′-dideoxy nucleoside triphosphates were tried. These analogs are incorporated in a sequence-specific manner by *E. coli* Pol I, but the enzyme is unable to further extend the growing DNA strand (in the case of the ddNTPs simply because of the lack of a 3′ hydroxyl group). Synthesis was carried out in the presence of all four dNTPs, one of which was α-^{32}P labelled. Four reactions were set up, each doped with a chain-terminating analog of one of the dNTPs, at an appropriate concentration. If the concentration of ddATP, for example, was adjusted so that it was incorporated in place of the normal dATP ~1% of the time then a series of chain-terminated products were produced, each ending with an A. Some molecules in the product ended at each of the A residues in the sequence. When such a product was electrophoresed on a denaturing 12% acrylamide gel, a series of bands representing the positions of all A's in the sequence were displayed. Unlike the plus and minus method, bands were produced for each A within runs of consecutive A residues. When the four dideoxy reactions were run in adjacent lanes it was possible to read sequences of ~100 nt in most cases. The dideoxy sequencing method as originally described required a single-stranded DNA template. The general applicability of the method was therefore greatly enhanced when Messing and collaborators developed methods for cloning into the single-stranded phage M13. During this period the useful read length of dideoxy sequencing increased from about 100 up to about 400. This improvement was mainly the result of (i) the use of very thin sequencing gels and (ii) ^{35}S labeling of the DNA, which gives sharper bands than ^{32}P due to the lower energy of the emitted ß particles. Sequencing capacity was also increased by the use of gels with narrow lanes, typically 48 lanes on a 20 by 45 cm gel. Sequencing reactions could be done manually in 96-well plates with handheld repetitive pipetting devices. During this period a single person could run 8 gels on a single day, each with 12 sequence ladders, and obtain some 30 kb of primary sequence data. But it was difficult to do this more than about twice a week (Fig. 2.2).

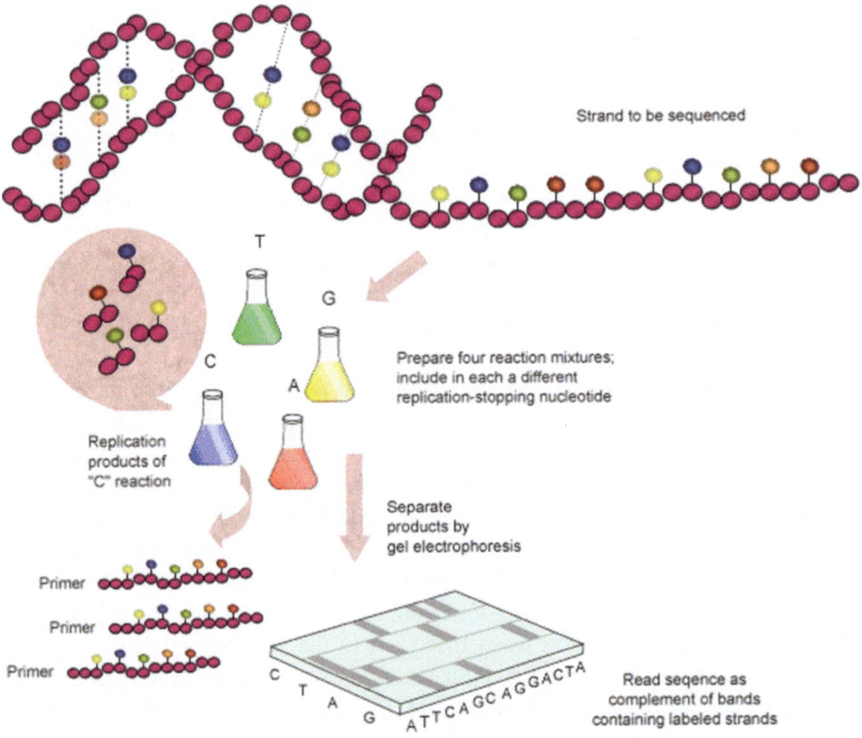

Fig. 2.2 The Sanger sequencing reaction. Single stranded DNA is amplified in the presence of modified radio-labelled dNTPs that terminate the reaction when they are incorporated into the nascent strand. The radio-labelled fragments produced are then separated using polyacrylamide gel electrophoresis

Bibliography

1. Bell, DA, DeMarini, DM. 1991. Excessive cycling converts PCR products to random-length higher molecular weight fragments. *Nucleic Acids Res.*, **19**, 5079.
2. Breslauer, KJ, Frank, R, Blbcker, H, Marky, LA. 1986. Predicting DNA duplex stability from the base sequence. *Proc. Natl. Acad. Sci. USA*, **83**, 3746–50.

3. Cheng, S. 1994. Long PCR. *Nature*, **369**, 684–5
4. Cheng, S. 1995. Template integrity is essential for PCR amplification of 20- to 30-kb sequences from genomic DNA. *PCR Methods Appl.*, **4** (5), 294–8
5. Cheng, S. 1994. Effective amplification of long targets from cloned inserts and human genomic DNA. *PNAS.*, **91**, 5695–9
6. Chou, Q, Russell, M, Birch, DE, Raymond, J, Bloch, W. 1992. Prevention of pre-PCR mis-priming and primer dimerization improves low-copy-number amplifications. *Nucleic Acids Res.*, **20**, 1717–23
7. Cobb, BD, Clarkson, JM. 1994. A simple prcedure for optimizing the polymerase chain reaction using modified Taguchi methods *Nucleic Acids Res.*, **22** (18), 3801–5
8. D'Aquila, RT, Bechtel, LJ, Videler, JA, Eron, JJ, Gorczyca, P, Kaplan, JC. 1991. Maximizing sensitivity and specificity of PCR preamplification heating. *Nucleic Acids Res.*, **19**, 3749
9. Demmler, GJ, Buffone, GJ, Schimbor, CM, May, RA. 1988. Detection of cytomegalovirus in urine from newborns by using polymerase chain reaction DNA amplification. *J. Infect. Dis.*, **158**, 1177–84.
10. Eckert, KA, Kunkel, TA. 1991. The fidelity of DNA polymerases used in the polymerase chain reactions. In PCR A Practical Approach; Eds. Mcpherson MJQuirke PTaylor GR, IRL Press, Oxford.
11. Erlich, HA, Gelfand, D, Sninsky, JJ. 1991. Recent advances in the polymerase chain reaction. *Science*, **252**, 1643–51
12. Foord, O. 1994. Long-distance PCR. *PCR Methods Appl.*, **3** (6), S149–S161
13. Fujiwara, H. 1995. PCR with deoxyinosine-containing primers using DNA polymerases with proofreading activity. *PCR Methods Appl.*, **4** (4), 239–40
14. Finckh, U, Lingenfelter, PA, Myerson, D. 1991. Producing single-stranded DNA probes with the Taq DNA polymerase: A high yield protocol. *BioTechniques*, **10**, 35–39
15. Gelfand, DH. 1989. TaqDNA polymerase. In PCR Technology Principles and Applications for DNA Amplification; Ed. Erlich HA, Stocktom Press, New York, NY.
16. Gelfand, DH, White, TJ. 1990. Thermostable DNA polymerases. In PCR protocols a guide to methods and applications; Eds. Innis, MAGelfand, DHSninsky, JJWhite, TJ, Academic Press, New York.
17. Her, C. 1995. Rapid Restriction Mapping by use of long PCR. *BioTechniques*, **19** (4), 530–2
18. Holland, PM, Abramson, RD, Watson, R, Gelfand, DH. 1991. Detection of specific polymerase chain reaction product by utilizing the 5′ -> 3′ exonuclease activity of Thermus aquaticus DNA polymerase. *Proc. Natl. Acad. Sci., USA*, **88**, 7276–80
19. Hoof van Huijsduijnen, RAM, Ayala, G., DeLamarter, F. 1992. A means to reduce the complexity of oligonucleotides encoding degenerate peptides. *Nucleic Acids Res.*, **20**, 919
20. Innis, MA, Gelfand, DH, Sninsky, JJ (eds), 1995. PCR Strategies. Academic Press, Inc, California, USA.
21. Innis, MA, Mayambo, KB, Gelfand, DH, Brow, MAD. 1988. DNA sequencing with Thermus aquaticus DNA polymerase and direct sequencing of polymerase chain reaction-amplified DNA. *Proc. Natl. Acad. Sci.*, USA, **85**, 9436–40
22. Jones, DH, Winistorfer, C. 1992. Sequence specific generation of a DNA panhandle permits PCR amplification of unknown flanking DNA. *Nucleic Acids Res.*, **20**, 595–600
23. Koike, K, Urata, Y, Koike, M. Human Pyruvate Dehydrogenase beta-subunit-gene, complete cds. Un published data, EMBL accession number D90086
24. Kwok, S. 1991. A guide to the design and use of mismatched and degenrate primers. *PCR Methods Appl.*, **3** (4), S39–S47
25. Kwok, S, Kellogg, DE, McKinney, N, Spasic, D, Goda, L, Levenson, C, Sninsky, JJ. 1990. Effects of primer-template mismatches on the polymerase chain reaction: human immunodeficiency virus type I model studies. *Nucleic Acids Res.*, **18**, 999–1005
26. Lawyer, FC, Stoffel, S, Saiki, RK, Mayambo, K, Drummond, R, Gelfand, DH. 1989. Isolation, characterization, and expression in Escherichia coli of the DNA polymerase gene from Thermus aquaticus. *J. Biol. Chem.*, **246**, 6427–37

27. McPherson T. PCR with highly degenerate primers PCR: A Practical approach. Ed. McPherson T. 1991, Oxford University, New York, pp. 171–86
28. Mertz LM. 1994. Nucleotide imbalance and PCR: Effects on DNA amplification and synthesis of high specific activity radiolabeled DNA probes. *Anal. Biochem.*, **221**, 160–65
29. Mezei, LM. 1990. Effect of oil overlay on PCR amplification. Amplifications a Forum for PCR Users, 4, 11–13
30. Mullis, KB. 1991. The polymerase chain reaction in an anemic mode: How to avoid cold oligodeoxyribonuclear fusion. *PCR Methods Appl.*, **1**, 1–4
31. Mullis, KB, Faloona, FA. 1987. Specific synthesis of DNA in vitro via a polymerase-catalyzed chain reaction. *Methods Enzymol.*, **155**, 335–50
32. Peterson MG, Tjian R. 1993. Cross-species polymerase chain reaction: Cloning of TATA box-binding proteins. *Methods Enzymol.*, 218, Anonymous Academic Press, 493–507.
33. Rappolee, DA. 1990. Optimizing the sensitivity of RT-PCR. Amplifications A Forum for PCR Users, 4, 5–7
34. Reischl, U, Wittwer, C, Cockerill, F. (eds), 2002. Rapid cycle real-time PCR. Methods and applications. Springer-Verlag Berlin Heidelberg, Germany.
35. Roux, K. 1994. Using mismarched primer-template pairs in touchdown PCR. *BioTechniques*, **16** (5), 812–4
36. Roux, K. 1995. Optimization and troubleshooting in PCR. *PCR Methods Appl.* **4** (5), S185–S194
37. Rolfs, A, Schuller, I, Finckh, U Weber-Rolfs, I. (eds), 1992. PCR: Clinical diagnostics and research. Springer-Verlag Berlin Heidelberg, Germany.
38. Ruano, G, Brash, DE, Kidd, KK. 1991. PCR: The first few cycles. Amplifications A Forum for PCR Users, 7, 1–4
39. Rychlik, W. 1995. Priming efficiency in PCR. *BioTechniques*, **18** (1), 84–90
40. Rychlik, W, Spencer, WJ, Rhoads, RE. 1990. Optimization of the annealing temperature for DNA amplification in vitro. *Nucleic Acids Res.*, **18**, 6409–12
41. Rychlik, W, Rhoads, RE. 1989. A computer program for choosing optimal oligonucleotides for filter hybridisation, sequencing and in vitro amplification of DNA. *Nucleic Acids Res.*, **17**, 8543–51
42. Saiki, RK, Gelfand, DH, Stoffel, S, Scharf, SJ, Higuchi, R, Horn, GT, Mullis, KB, Erlich, HA. 1988. Primer-directed enzymatic amplification of DNA with a thermostable DNA polymerase. *Science*, **239**, 487–91.
43. Saiki, RK, Scharf, S, Faloona, F, Mullis, KB, Horn, GT, Erlich, HA, Arnheim, N. 1985. Enzymatic amplification of ß-globin genomic sequences and restriction site analysis for diagnosis of sickle cell anemia. *Science*, **230**, 1350–54
44. Sambrook, J, Fritsch, EF, Maniatis, T. 1989. Molecular cloning. A laboratory manual, 2.Edition, CSH
45. Shibata, D, Martin, WJ, Appleman, MD, Causey, DM, Leedom, JM, Arnheim, N. 1988. Detection of cytomegalovirus DNA in peripheral blood of patients infected with human immunodeficiency virus. *J. Infect. Dis.*, **158**, 1185–92
46. Sommer, R, Tautz, D. 1989. Minimal homology requirements for PCR primers. *Nucleic Acids Res.*, **16**, 6749
47. Steffen, C. (ed), 1999. PCR Applications Manual, 2nd edition. Roche Diagnostics GmbH, Mannheim, Germany.
48. Taylor, GR. 1991. Polymerase chain reaction: basic principles and automation. In PCR a practical approach; Eds. Mcpherson, MJ, Quirke, P, Taylor, GR, IRL Press, Oxford
49. Thein, SL, Wallace, RB. 1986. The use of synthetic oligonucleotides as specific hybridisation probes in the diagnosis of genetic disorders. In Human genetic diseases: a practical approach. Eds. Davis K.E. IRL Press, Herndon, Virginia
50. Tindall, KR, Kunkel, TA. 1988. Fidelity of DNA synthesis by the Thermus aquaticus DNA polymerase. *Biochemistry*, **27**, 6008–13

51. Walsh, P. 1992. Preferential PCR Amplification of Alleles: Mechanisms and Solutions. *PCR Methods Appl.*, **1** (4), 241–50
52. Yap, EPH, McGee, OD. 1991. Short PCR product yields improved by lower denaturation temperatures. *Nucleic Acids Res.*, **19**, 1713
53. White, BA. (ed), 1993. PCR protocols. In Methods in Molecular Biology, volume 15. Humana Press, New Jersey

Chapter 3
Real-Time PCR – The Basic Principles

Contents

3.1 Traditional PCR Versus Real Time PCR

Conventional PCR is a powerful technique that allows exponential amplification of DNA sequences. A PCR reaction needs a pair of primers that are complementary to the sequence of interest. Primers are extended by the DNA polymerase. The copies produced after the extension, so called amplicons, are re-amplified with the same primers leading thus to an exponential amplification of the DNA molecules. After amplification, gel electrophoresis is used to analyse the amplified PCR products and this makes conventional PCR time consuming; since the reaction must finish before proceeding with the post-PCR analysis. Real Time PCR overcome this problem, because of its ability to measure the PCR amplicons at early states of the reaction as they are accumulate in a "Real Time Detection" mode thus measuring the amount of PCR product where the reaction is still in the exponential phase (QPCR).

E.A. Pestana et al., *Early, Rapid and Sensitive Veterinary Molecular Diagnostics - Real Time PCR Applications*, DOI 10.1007/978-90-481-3132-7_3,

Real Time PCR allows detection and quantitative measurement of products generated during each cycle of the PCR process that are directly proportional to the amount of the template DNA before the start of the PCR process. Such chemistry requires the use of a method to detect the product formed on each cycle and of a thermocycler that is adapted to record the results obtained on each amplification cycle in a Real Time manner.

3.1.1 PCR Kinetics

Differences between conventional PCR and QPCR are much easier to understand when the kinetics of the PCR reaction are examined. Let us imagine we have three replicates. As such all samples will begin the PCR cycling process with the same conditions: the same quantity of all PCR components and the same DNA concentration.

A conventional PCR reaction usually steps into 3 phases: the *exponential*, the non-exponential and *plateau* or *end-point* phase (Fig. 3.1).

At the beginning of each PCR reaction, all components are present in a sufficiently high quantity to guarantee good amplification and as the PCR progresses and fresh components are present, amplification occurs in an *exponential* manner, that is the reaction proceed doubling the quantity of initial DNA with every other cycle. As the cycles progress and reagent components of the reaction start to be depleted, the reaction will begin to slow down and the PCR product will no longer be doubled in every cycle, and the non-exponential *amplification* occurs where samples begin to diverge in their quantities. After several rounds of amplification, the PCR reaction will no longer generate template due to the lack of critical components in the reaction, what it is commonly known as the plateau phase or end-point of the PCR reaction.

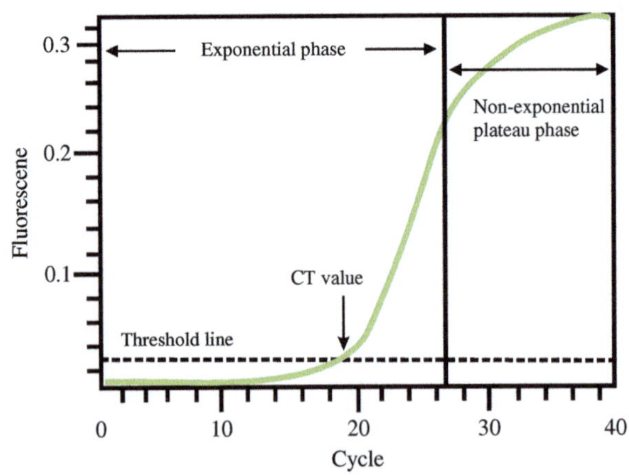

Fig. 3.1 Amplification plot. Baseline fluorescence is subtracted

All samples that started with the same quantity of all PCR components and DNA concentration at the beginning of the reaction will end up at different points in the plateau phase, due to the different kinetics each sample had during cycling. Therefore, it is more precise to take measurements during the exponential phase, where the replicates samples are amplifying exponentially. In conventional PCR, the products are analyzed by size-fractionation of the amplicons by gel electrophoresis and ethidium bromide. This process therefore will measure the final DNA obtained during the PCR reaction, which can be compared to measuring the DNA in the plateau phase (terminal phase). At this point, quantification is exceedingly difficult since the PCR gives essentially the same amount of product independently of the initial amount of DNA template molecules that were present.

Note that End-point detection is very time consuming since we must wait until the PCR reaction is finished to load it onto a gel for quantification. In this sense, Real time Quantitative PCR is also more effective than conventional PCR since no post-PCR processing of PCR products is needed, and thus chances of carryover contamination are reduced and efficiency is achieved in a high throughput manner.

Since its invention, PCR has evolved from being monitored on gel electrophoresis, to a reliable and easy to used method for detection and quantification of nucleic acid sequences. Real time PCR or also called quantitative PCR (QPCR) or kinetic PCR, allows PCR product accumulation being measured via different fluorescent chemistries that arises during the amplification process. Such chemistries will be analyzed in detail later on.

3.2 Optimising a Real-Time PCR Reaction

3.2.1 Primer Sets and Probe Design

The first step in designing a primer is to obtain the sequence of the gene of interest. Several numbers of public databases, such as the NCBI, offer help to achieve that. After the sequence is obtained, software to design primers is used to simplify and maximize the success for the design process. Softwares are coupled with the rules making this process simple and reliable. Software programs can be easily and freely obtained from Internet or through most oligonucleotide vendors.

The next step is to consider carefully the area of the gene to be used as the detection template. Regions with secondary structures or long repetitive nucleotides should be avoided. When coding regions are considered as targets, it is advisable to design primers that allow differentiation between the amplified product of cDNA and an amplified product derived from contaminating genomic DNA.

There are two approaches on designing the required primers. The first one is to design primers that anneal to sequences in exons on both sides of an intron. With such primers, any product amplified from genomic DNA will be much larger than a product amplified from intronless mRNA. The second alternative is to make primers that span exon/exon boundaries on the mRNA, in that way amplification of genomic

DNA is avoided. Chemical treatment of the sample with DNAse prior to the RT step is also a good approach.

Primers and probes should be selected in a region with G/C contents of 20–80%. Lower G/C content is preferred. Generally, higher performance is achieved when the region spanned by primers is between 50 to 150 bp in length for probe-based chemistries, and between 100–400 bp in length if SYBR® Green I is used. When using probe based chemistries, the target for amplification should be keep as short as possible to maximize the efficiency of the assay. In the case of SYBR® Green I, which binds to double strand, shorter targets are also advantageous in terms of an efficient amplification, however, longer fragments will produced higher fluorescent signal, due to more dye molecules binding and will improve detection sensitivity.

As in a conventional PCR set up, primers should be between 15 to 30 bp in length, and the theoretical T_m between the primers should not exceed 2 degrees between them. It is also advisable to avoid any secondary structures such as G/C clamps at the 3′end of the primers to avoid self-folding or non-specific annealing. The five bases at the 5′ terminal end should not contain more than two guanines and cytosines, although it is acceptable to have 3 in the final 5 bases if no two pyrimidines are adjacent. Since thymidine tends to miss-prime more readily than the other bases, a 3′ terminal T should be avoided if possible. The 5′ end of the primers should also not contain an inverted repeat sequence that would allow it to fold on itself. Finally, it is important to check all primers and probes to make sure they do not anneal with other targets, and thus to avoid false positive results.

3.2.1.1 Primer Secondary Structures

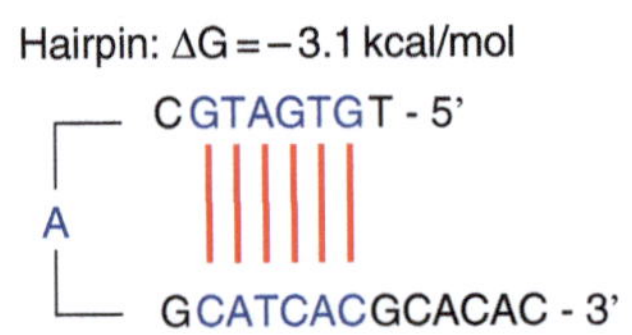

Hairpins are formed by intra-molecular interaction within the primer. These structures should be avoided when designing Real-Time PCR primers.

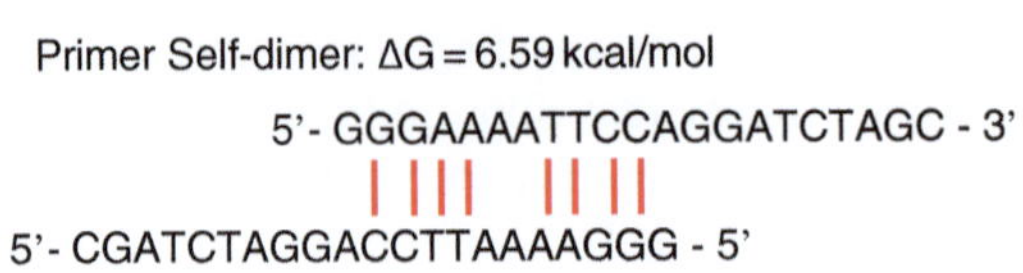

Self dimers are formed by intermolecular interactions between the two (same sense) primers, where the primer is homologous to itself. When this structures are present, large amount of primers are needed in PCR, and thus the product yield is reduce.

Cross-dimer: ΔG = 6.59

```
                 5'-TGTGATGCAGCATCACGCACAC -
                    || ||      ||
3' - CTACGTCGACTCTGATAGCTACG -
```

Cross dimers are formed by intermolecular interaction between sense and antisense primers, where they are homologous.

3.2.2 PCR Components and Assay Optimization

Real Time PCR optimization comprises almost the same steps one should follow when optimizing a conventional PCR reaction. The variations are according to the chemistry selected to detect the newly amplified DNA. Detailed information for optimisation of each of the components is detailed below. Starting with the most important component for the successfulness of the QPCR reaction, the steps usually follow for standardisation are:

- Optimization of primers sets and probes, Identification of the correct concentration and evaluation of possible secondary structures (SYBR® green I).
- Magnesium chloride and nucleotide concentration.
- Assay efficiency: Analysis of the slope of a DNA standard dilution series.
- Assay precision: Replicate number (Triplicates are always recommended).

There are a number of companies that offer QPCR master mixes, but PCR reagents can be also purchased separately and in this case optimisation plays an important role on the success of the reaction. PCR components can be bought from any commercial company. Taq Polymerase however has to be purchase to meet an efficiency and speed of amplification, and according to the type of chemistry to be used on the experiment, for example if a TaqMan probe is used as detection chemistry, then a polymerase with $5'$ to $3'$ exonuclease activity is required. Usually proofreading enzymes are not needed, and should not be used with probe base chemistry because they use a $3'$ to $5'$ exonuclease activity to eliminate miss-incorporated bases, and thus they can digest the probe.

To start with the validation and optimisation, the first element to be taken into consideration is a good starting control template. Most commonly purified PCR fragments, genomic DNA, or cDNAs can be used as controls. Once a control source has been identified, it is time to play with the rest of the PCR components.

The first component that needs to be evaluated is the PCR primer concentration. Primer concentrations need to be adjusted according to the template used on the reaction, and the chemistry being used on each assay. For SYBR® green I, usually low primer concentrations are needed to avoid primer-dimer formation. Most SYBR® green assays use primer concentrations from 50 to 300 nM. For Fluorescent resonance energy transfer (FRET) probes, such TaqMan or Molecular Beacons, ranges from 50 to 600 nM should be enough for standardisation of the probe.

Primer optimisation can be easily achieved by the use of SYBR® green I. SYBR® green I is a double strand binding dye. When SYBR® green I is added to a QPCR reaction, the generation of specific and unspecific products is visualized on the melting curve formed after the amplification is finished. As a result, SYBR® green I can be used as a tool to not only to determine primer performance, but also primer specificity at different concentrations. When performing an assay with TaqMan technology, primer pairs should be design to contain no secondary structures that can jeopardize the reaction later. In this sense SYBR® green I can help to determine primer performance before ordering the probe. This is an important asset since redesigning the primers would also imply the redesigning of the probe, and therefore running this assay prior ordering the probe is highly recommended.

When analyzing different primer combinations, the primer concentration indicating the lowest Ct values should be chosen. Melting curves should generate a single peak of PCR product. The overall performance of the QPCR is determined by the use of serial dilutions of a template control, procedure known as standard curve. Standard curves help to determine the efficiency, accuracy and sensitivity of the primers and probes. Standard curves are generated diluting the template control into two to ten fold dilution series. Dilution series in triplicates are recommended to evaluate reproducibility and possible pipetting errors.

After QPCR amplification, efficiency is calculated by using the slope of the standard curve. Thriving efficiency is achieved when the amount of template control is doubling with each cycle. The standard curve plot of the log of starting template vs. PCR cycles should also generate a linear fit with a slope between –3.1 and –3.6, thus this results indicating reaction efficiency of 96–110%.

If, after primer and probe standardization the assay is still not performing well, magnesium chloride concentrations can be adjusted. Usually magnesium chloride concentrations within a range of 3.5–5.5 mM are suitable for TaqMan and Molecular Beacons assays. For SYBR® green I 1.5–3.5 mM range of Magnesium chloride is recommended. If the reaction is still not performing as desired, primers are probes should be redesign.

3.2.3 Real-Time Fluorescense Reporters

Most commonly used chemistries available for QPCR are SYBR® Green I, TaqMan, Molecular Beacons, and Scorpions. The chose of a particular chemistry will depend on the experimental design and objectives of the experiment. Each fluorescence reporter has its own advantages and shortcomings. To simplify the explanation of each technology, we will divide the available chemistry into two groups. The DNA binding dyes, and the Probe-based chemistries. All these chemistries will allow the detection of newly PCR products formed via the fluoresce signal generated. Classical intercalators such as SYBR® Green I are non-specific labelers; as they bind to any double-stranded DNA. Sequence specific probes such as TaqMan, Molecular Beacons and Scorpions use the presence of a Fluorescence resonance energy transfer (FRET) probe as a reporter system, and the 5′ exonuclease activity of the DNA polymerase to detect PCR amplification in Real-Time. We will examine all of them in detail.

3.2.3.1 DNA Binding Dyes, SYBR® Green I

Advantages

The fluorescence intensity measured is proportionate to the amount of PCR products produced.
Can monitor the amplification of any dsDNA
Since no specific probe is required, the set-up time and costs are reduced

Disadvantages

Is a non-specific dye, binds to any dsDNA, therefore can generate false positive signals coming from primer-dimer formation or non-specific products
Requires the analysis of the melting curve at the end of the amplification reaction to check for possible primer-dimer or non-specific product formation

SYBR® green I is a cost effective dye, compared to other chemistry's, and easy to use, which allows the initial optimization of primers for any QPCR chemistry. When SYBR® green I is free in solution displays relatively low fluorescence levels, but when dsDNA is added, the fluorescence increases over 1000-fold. As the accumulation of newly PCR amplicons increases with the reaction cycles, the intensity of fluorescence also increases as more SYBR® green I molecules are binding to the dsDNA, and thus the accumulation of product can be measured on real time (Fig. 3.2).

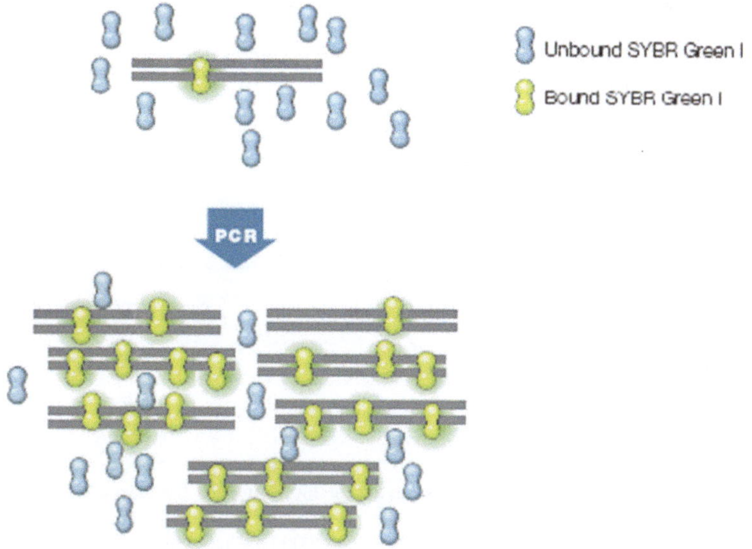

Fig. 3.2 DNA binding dyes in Real time PCR. Fluorescence dramatically increases when the dye molecule binds to DNA. (Courtesy of Bio-Rad Laboratories, Inc.)

SYBR® green I assays depend exclusively of the designed primers. If the primer pairs are designed not to produce non-specific-binding, secondary structures or dimers, then optimization of the amplification efficiency and specificity are straightforward steps. As sometimes non-specific signal cannot be avoided, it is important after every assay to analyse the melting curve. The ability of the SYBR® green I to bind dsDNA, and be free in the reaction as ssDNA is formed, it is used to allow the formation of the melting curve. During denaturation, PCR products are melted (ssDNA) to allow the polymerase to anneal. During this step as the temperature rises, the fluorescence recorded will also decrease. The melting temperature (T_m) depends in part of the base composition of the product formed. All PCR products for a particular primer set should have the same melting temperature, unless there is a contamination, mispriming, or primer-dimer artefacts. As mentioned before, SYBR® green I has not the capability to distinguish between one DNA and another, and thus observing similar melting curves at the end of the PCR is important. If the PCR molecules formed are homogeneous in length, a single thermal transition should be detected (a single curve). On the contrary, if more than one of the PCR products formed are not homogeneous; multiple thermal transitions will be observed (Fig. 3.3).

After Real-Time PCR is finished, the thermocycler can be adapted to perform a melting curve by means of increasing the temperature by a fraction of a degree and the measuring of the change in fluorescence. At the melting point, the two strands of DNA separate and the signal fluorescence decreases. The software plots the rate of change of the relative fluorescence units (RFU) with the time (T)

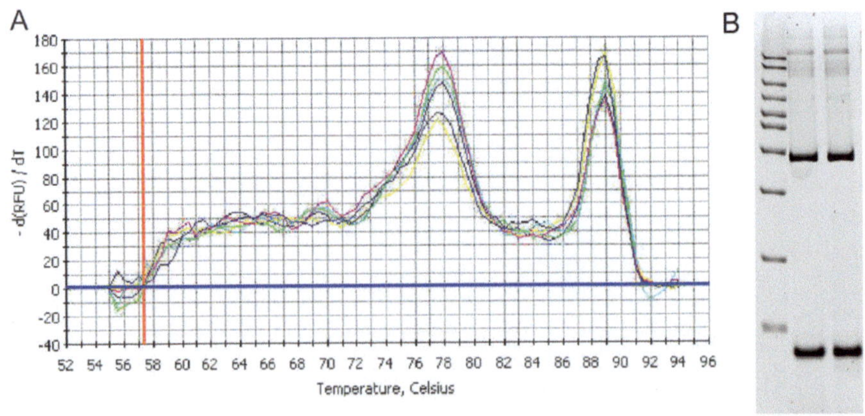

Fig. 3.3 Melt-curve analysis of reaction product from a SYBR® Green assay. The melt curve analysis function of real-time instruments can be used to distinguish specific products from non-specific products. **A**. The negative first derivative of the change in fluorescence is plotted as a function of temperature. The two peaks indicate the T_m values of the two PCR products. **B**. Gel analysis of the PCR products. Lane 1, AmpliSize 50–2000 base pairs (bp) molecular ruler. Lanes 2 and 3, two replicates of PCR product from the reaction shown in A. The two PCR products are revealed as separate bands in the gel. (Courtesy of Bio-Rad)

(–d(RFU)/dT) on the Y-axis versus the temperature of the X-axis, and this will peak at the melting temperature (T_m). The sensitivity of detection with SYBR®-Green I may be compromised by the formation of primer-dimers, lack of specificity of the primers, primer concentration (which can be limiting) and the formation of secondary structures in the PCR product. All of these factors could lead to the creation of unexpected double-stranded DNA product, which would incorporate SYBR®-Green I and register a fluorescent signal. For example, if a curved is formed at the left side of the main melting curve, this, in most cases indicates the presence of primer-dimmer artefacts, giving a peak with lower melting temperature, as the formed artefact is a very short DNA.

Optimisation of the primer concentration is therefore essential. Primer concentration is usually determined to be optimal when specific amplification relative to primer-dimers is maximal in a positive versus negative control experiment. The primers should be selected to give a product (amplicon) of 100–400 bp. SYBR® Green dye fluorescence depends on the presence of double-stranded DNA, therefore products should not be too short. If necessary, primers that generate amplicons as small as 100 bp may also be acceptable. The concentration of the primers should not be so high; usually it is advisable to start with low primer concentrations to avoid both mispriming and dimerization. Template regions with obvious secondary structures or long runs of the same nucleotide should be avoided when designing primers. It is advisable to avoid a 3′ terminal T on primers if possible. The reason for this is that thymidine tends to misprime more readily than other bases, and thus annealing of the primer at other places on the genome can occur giving race to non-specific amplification. SYBR® green is usually used for optimization of primers prior to ordering the sequence specific probe. As the probe is expensive, it is advisable to always check the efficiency of the primers.

3.2.4 Melting Curve Dissociation Analysis

During the PCR reaction, amplicons that are being formed are going through cycles of denaturaturation (usually at 95°C) and annealing (usually around 50–55°C), and they are subject to gradual increases in temperature to give increases I amplicon through denaturation, These gradual increases are monitored as changes in fluorescence (as SYBR® green will intercalate at 55°C and dissociate at 95°C) and are recorded and plotted against the temperature.

When analysing a melting curve one should focus on two aspects, the presence of a single melting product and in evaluating the NTC sample for primer dimmer formation. If these two parameters are reliable, then the CTs obtained can be trusted as accurate.

The SYBR® Green I dye chemistry can be used for the following assay types:

- One-step RT-PCR for RNA quantitation
- Two-step RT-PCR for RNA quantitation
- DNA/cDNA quantitation

3.2.5 Probe-Based Chemistry

Probe-based chemistries are more specific for detection due to the fact that they use an internal probe besides the pair of primers used to amplify the region of interest. All probes rely on the close proximity of a quenching dye to the reporter dye. The effect is called Foerster Resonance Transfer or Fluorescence Resonance Transfer (FRET). In most cases disruption of this quenching interaction causes an increase in fluorescence that is proportional to product formation. However, FRET probes rely on the formation of a quenching interaction and quantification relies on measuring an increase in fluorescent signal ongoing with PCR product amplification. Compared to SYBR® Green, all probes are more specific, but they are more expensive and eventually require more validation and optimization.

3.2.5.1 FRET-based Hydrolysis Probes, TaqMan Probes

Advantages of TaqMan Chemistry

- Specific hybridization between probe and target is required to generate fluorescent signal.
- Probes can be labelled with different, distinguishable reporter dyes, which allows amplification of multiple distinct sequences in one reaction tube.
- Post-PCR processing is eliminated, which reduces assay labor and material costs.

Disadvantages of TaqMan Chemistry

- TaqMan chemistry requires the synthesis of different probes for different sequences, making the assay specific, but at the same time expensive.
- Probe-based assay design can be more challenging than SYBR green assay set up.

TaqMan probes consist of a single-stranded oligonucleotide that is complementary to a sequence within the target template. The probe has a fluorescent dye at its 5′ end, whose signal is silenced by a quencher dye molecule, e.g. the Black Hole Quencher (BHQ), at the 3′ end. TaqMan probes use a FRET (Fluorescence Resonance Energy Transfer) as a quenching device, so as long as the quencher and the fluorescent dye are on close proximity, quenching will occur. Soon after the TaqMan probe hybridizes to one of the strands on the template, it is digested by the 5′ exonuclease activity of the *Taq* DNA polymerase as it extends the amplification primers. Cleaving of the probe releases the fluorescent dye from the quencher, resulting in an irreversible increase in the fluorescence signal. Fluorescence will increase as the PCR cycles progress, proportional to the rate of probe cleavage and ideally proportional to the rate of amplified DNA.

3.2.5.2 TaqMan Probe Design Guidelines

1. Probes should not contain more than 3 consecutive runs of the same base, and should contain more C than G nucleotides, due to the strong interaction that G displays with itself. Guanine is a very effective fluorescent quencher and therefore, should not be adjacent to the reporter dye.
2. TaqMan probes should be place as close as possible towards the 5′ end of the amplicon. The reason for this is that the closer the probe to the 5′ end, the sooner the probe will be cleaved thus giving an immediate signal.
3. The size of the probe should be between 20 and 30 bp in length, and have balance GC content. T_m requirements of the probe are most often dictating the specific %GC; TaqMan probe T_m and the T_m targets without the primers annealing at the same time, so this step is performed at a temperature too high for the primers to anneal. When the temperature of the reaction is at the anneal step, the primers will extend and cleave off the probe (Fig. 3.4).

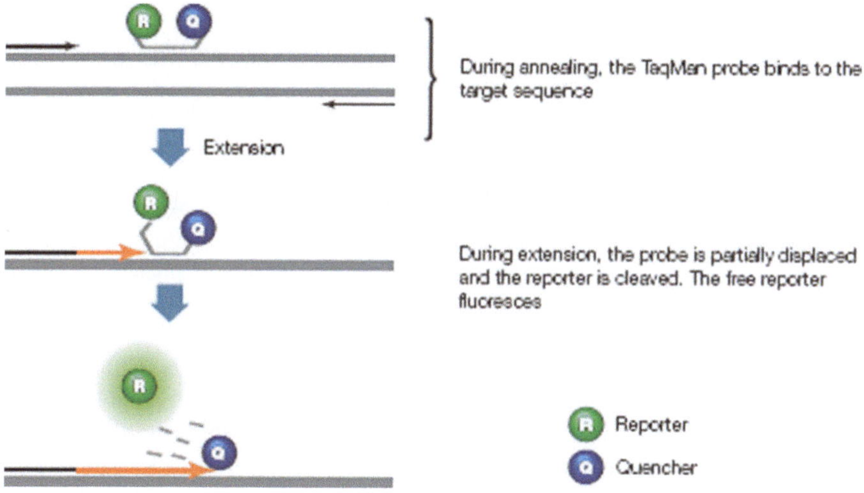

Fig. 3.4 TaqMan Assay (Courtesy of Bio-Rad)

The advantage of using a TaqMan probe is the high specificity, the high signal to noise ratio and the ability to detect multiple sequences in one reaction. This might be of impact and quite benefitial when the DNA material is very rare.

3.2.5.3 MGB TaqMan Probe

This is a variant of the original TaqMan chemistry, with the addition of a Minor Groove Binder (MGB) at the 3′ end that stabilise the T_m of the probe, and thus a shorter probe can be used (from 30 down to 13 bp), which makes them better suited for allelic discrimination applications so called SNPs (single nucleotide polymorphism).

The TaqMan chemistry can be used for the following assay types:
Quantitation, including:
One-step RT-PCR for RNA quantitation
Two-step RT-PCR for RNA quantitation
DNA/cDNA quantitation
Allele discrimination: SNP genotyping assays

3.2.6 FRET-Based Hybridisation Probes

3.2.6.1 Molecular Beacons

Advantages

Lower background, greater specificity

Disadvantages

More difficult to design and optimize

Molecular beacons are single-stranded oligonucleotide hybridization probes that form a stem and a loop structure. The loop contains a probe sequence that is complementary to a target sequence, similar to the TaqMan probe. The stem structure is formed by addition of 5–6 bases forming complementary arms that are located on either side of the probe sequence, and are annealing to each other. A fluorophore is covalently linked to the end of one arm and a quencher is covalently linked to the end of the other arm. The stem keeps these two moieties in close proximity to each other, causing the fluorescence of the fluorophore to be quenched by energy transfer.

Since the quencher moiety is a non-fluorescent chromophore and emits the energy that it receives from the fluorophore as heat, the probe is unable to fluoresce. When the probe encounters a target molecule, it forms a hybrid that is longer and more stable than the stem and its rigidity and length preclude the simultaneous existence of the stem hybrid. Thus, the molecular beacon undergoes a spontaneous conformational reorganization that forces the stem apart, and causes the fluorophore and the quencher to move away from each other, leading to the restoration of fluorescence.

Molecular Beacons do not rely on the exonuclease activity of the Taq polymerase, and so traditional 3 steps PCR is suitable for this purpose. During the extension step at 72°C, the Taq polymerase will extend the sequence to where the molecular beacon probe is annealed, displacing the probe, and so once in solution the probe will reassume its stem-loop conformation.

When designing the probe two issues should be carefully considered. The stem structure should not be too stable, otherwise target hybridization can be inhibited. The stem-loop structure should always be refolded in the same stem-loop conformation after displacement. When this process does not occur, the fluorochome and quencher will not quench proper, and signal is always observed (Fig. 3.5).

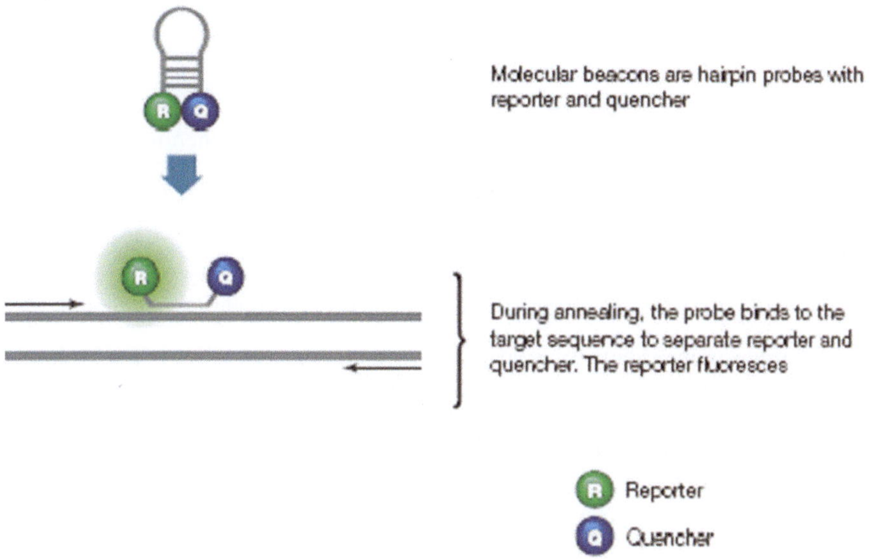

Fig. 3.5 Molecular Beacons (Courtesy of Bio-Rad)

3.2.6.2 Probe Design

Molecular Beacons are design to anneal at 7–10°C higher than the primers, to allow hybridization before the primers are extended.

The stem sequence should be of about 5–7 bp long and should have a similar T_m to the melting temperature of the probe region in the loop. There is a general consensus about the size of the stem sequence and the T_m that will be generated. Usually, 5 bp long stem will have a T_m of 55–60°C, 6 bp long stems have a 60–65°C T_m, and 7 bp long stems will have a 65–70°C T_m. It is also advisable to check the correct folding or stem-loop structure of the Beacon in a simulation folding program like "mfold". Typically, the Beacon is designed to anneal in the middle of the target sequence (which is the region comprise between the two primers). This is important to guarantee that in an event of any low-activity extension by the Taq polymerase, during the annealing step, will not displace the probe before the fluorescence reading is taken.

Molecular Beacons can be used for the following assay types:

- SNP analysis
- Real-time Nucleic acid detection
- Real-time PCR quantification
- Allelic discrimination
- Multiplex assays
- Diagnostic clinical assays

3.2.7 *Scorpion Primers*

Advantages

Lower background, greater specificity.

Scorpions do not depend upon enzymatic cleavage and, therefore, rapid PCR cycling is possible.

Disdvantages

Much more difficult than TaqMan or Molecular Beacons to design and optimize.

Scorpion-probe chemistry is very similar to the ones described in the Molecular Beacon section, they consist of a stem-loop or hairpin loop configuration when they are not hybridised. The stem sequences that are keeping the hairpin loop configuration are located on the 5′ and 3′ sides of the probe. Additionally, a fluorophore is attached to the 5′-end and is quenched by a moiety attached to the 3′-end of the loop. The loop contains the specific probe sequence. The difference between Beacons and Scorpions lies on that there is no need for a separate probe. Scorpions incorporate 2 new elements: a PCR primer and a non-amplifier monomer or PCR stopper which prevents PCR read-through of the probe element. So all the system needed for detection (primer and probe) are contained in one single element.

After extension of the Scorpion primer, the specific probe sequence is able to bind to its complement within the same strand of DNA. This hybridization event opens the hairpin loop so that fluorescence is no longer quenched and an increase in signal is observed. The non-amplifier monomer prevents read-through, which could lead to opening of the hairpin loop in the absence of the specific target sequence. Such read-through would lead to the detection of non-specific PCR products, e.g. primer dimers or mispriming events.

Since the annealing of the loop sequence with the PCR product is an intramolecular interaction, it is kinetically more favorable than probe systems that contain two separated molecules that interact (probe and template). This is the reason why Scorpion usually result in higher fluorescence signal when compared to TaqMan and Molecular Beacons. Since Scorpions don't use the 5′-3′ exonuclease activity, a three step PCR cycling protocol is usually preferred.

3.2.7.1 Probe Design Specifications

Scorpion probes should be of approximately 17–30 bp length and it is advisable to place the probe no more than 11 bp upstream of the complementary target sequence. Usually when the complementary sequence is located too far away downstream, the probe loses efficiency. The stem sequence is recommended to be of about 6–7 bp in length, and to contain enough pyrimidines to allow the T_m of the stem loop structure to be 5–10°C higher than the T_m of the primer sequence to the target. Moreover, this

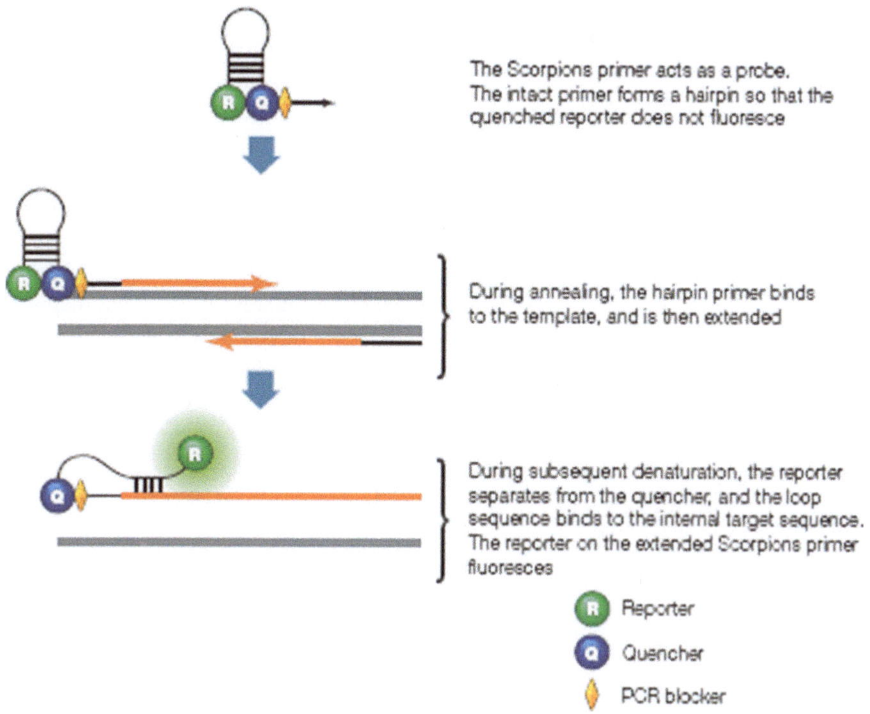

Fig. 3.6 Scorpion probes (Courtesy of Bio-Rad)

will allow the Gibb's free energy (*G*) to be negative enough to allow proper folding of the stem loop (Fig. 3.6).

Scorpions can be used for the following assay types:

QPCR (Quantitative Polymerase Chain Reaction): As explained above, the system consists of one Scorpion primer containing a primer and probe linked to each other and one compatible primer to target the other strand of the target are designed simultaneously.

ASO (Allele Specific Oligonucleotide) Scorpion: ASO-Scorpions are designed to discriminate between two alleles. Two Scorpions, one for each allele, and one compatible primer to target the other strand of the target of both alleles are designed simultaneously. The primer part of both Scorpions is identical; however the probe differs by one nucleotide, so that one will only bind to one of the alleles and the other only to the other allele.

ARMS (Amplification Refractory Mutation System) Scorpion: ARMS Scorpions are also used to amplify and discriminate between two possible alleles. Once again, two Scorpion primers are designed, one for each allele, and one compatible primer to target the other strand of both alleles are being designed (common primer). The probe parts of both Scorpions are identical; and the primer parts of the Scorpions are used to discriminate between the two alleles.

3.2.8 *LAMP*

"LAMP" stands for Loop-mediated Isothermal Amplification, is a simple, rapid, specific and cost-effective nucleic acid amplification method solely developed by Eiken Chemical Co., Ltd. LAMP technology uses 4 different primers particularly designed to recognize 6 distinct regions on the target gene and the reaction process proceeds at a constant temperature using strand displacement reaction.

Amplification and detection of a gene can be completed in a single step, by incubating the mixture of samples, primers, DNA polymerase with strand displacement activity and substrates at a constant temperature (about 65°C). It provides high amplification efficiency, with DNA being amplified 109–1010 times in 15–60 min. Because of its high specificity, the presence of amplified product can indicate the presence of the target gene.

Advantages of LAMP

- There is no need for a step to denature double stranded into a single stranded form.
- The whole amplification reaction takes place continuously under isothermal conditions.
- The amplification efficiency is extremely high.
- By designing 4 primers to recognize 6 distinct regions, the LAMP method is able to specifically amplify the target gene.
- The total cost can be reduced, as LAMP dose not require special reagents or sophisticated equipments.
- The amplified products have a structure consisting of alternately inverted repeats of the target sequence on the same strand.
- Amplification can be done with RNA templates following the same procedure as with DNA templates, simply through the addition of reverse transcriptase

3.2.8.1 Primer Design

Four types of primers are design based on the 6 distinct regions of the target gene: the F3c, F2c and F1c regions at the 3′ side and the B1, B2 and B3 regions at the 5′ side.

FIP: Forward Inner Primer (FIP) consists of the F2 region (at the 3′ end) that is complementary to the F2c region, and the same sequence as the F1c region at the 5′ end.

F3 Primer: Forward Outer Primer consists of the F3 region that is complementary to the F3c region.

BIP: Backward Inner Primer (BIP) consists of the B2 region (at the 3′ end) that is complementary to the B2c region, and the same sequence as the B1c region at the 5′ end.

B3 Primer: Backward Outer Primer consists of the B3 region that is complementary to the B3c region.

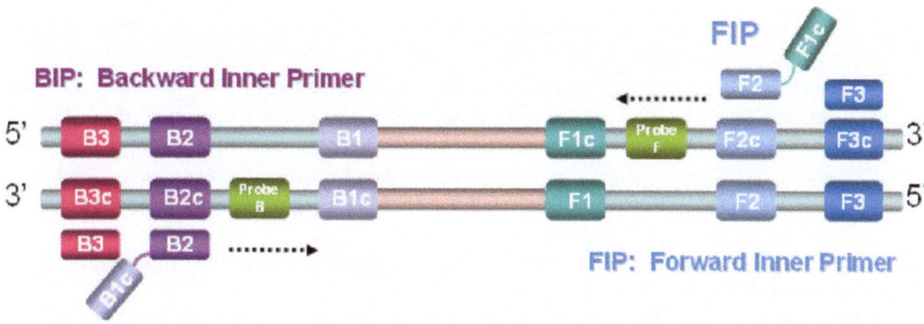

3.2.8.2 Primer Design Specifications

- The distance between 5′ end of F2 and B2 is considered to be 120–180 bp, and the distance between F2 and F3 as well as B2 and B3 is 0–20 bp.
- The distance for loop forming regions (5′ of F2 to 3′ of F1, 5′ of B2 to 3′ of B1) is 40–60 bp.
- The T_m of the primers should be of about 60–65°C in the case of GC rich and about 55–60°C for AT rich.
- G-C content should be of about 50–60% in the case of GC rich and Normal, about 40–50% for AT rich.
- Primers should be designed so as not to easily form secondary structures. 3′ end sequence should not be AT rich or complementary to other primers.

Bibliography

1. Heid, CA, Stevens, J, Livak, KJ, Williams, PM. 1996. Real time quantitative PCR . *Genome Res.*, **10**, 986–94.
2. Nolan, T, Hands, RE, Bustin, SA. 2006. Quantification of mRNA using real-time RT-PCR. *Nat. Protoc.*, **1**, 1559–82.
3. Mhlanga, MM, Malmberg, L. 2001. *Methods*, **25**(4), 463–71.
4. Chandler, DP, Wagnon, CA, Bolton, H Jr. 1998 Reverse transcriptase (RT) inhibition of PCR at low concentrations of template and its implications for quantitative RT-PCR. *Appl. Environ. Microbiol.*, **64**(2), 669–77.
5. de Silva, D, Wittwer, CT. 2000. Monitoring hybridization during polymerase chain reaction. *J. Chromatog. Biomed. Sci Appl.*, **741**(1), 3–13.
6. de Kok, JB, Hendriks, JC, van Solinge, WW, Willems, HL, Mensink, EJ, Swinkels, DW. 1998. Use of real-time quantitative PCR to compare DNA isolation methods. *Clin. Chem.*, **44**(10), 2201–04.

7. Boulay, JL, Reuter, J, Ritschard, R, Terracciano, L, Herrmann, R, Rochlitz, C. 1999. Gene dosage by quantitative real-time PCR . *Biotechniques*, **27**(2), 228–30.
8. Ferré, F. 1992. Quantitative or semi-quantitative PCR: Reality versus Myth. *PCR Methods Appl.* **2**, 1–9.
9. Higuchi, R, Fockler, C, Dollinger, G, Watson, R. 1993. Kinetic PCR analysis: real-time monitoring of DNA amplification reactions. *Biotechnology*, **11**, 1026–30.
10. Jung, R, Soondrum, K, Neumaier, M. 2000. Quantitative PCR. *Clin. Chem. Lab. Med.*, **38**(9), 833–36.
11. Kainz P. 2000. The PCR plateau phase – towards an understanding of its limitations. *Biochim. Biophys. Acta.*, **1494**(1–2), 23–27.
12. Mannhalter, C, Koizar, D, Mitterbauer, G. 2000. Evaluation of RNA isolation methods and reference genes for RT-PCR analyses of rare target RNA. *Clin. Chem. Lab. Med.*, **38**, 171–77.
13. Morrison, TB, Weis, JJ, Wittwer, CT. 1998. Quantification of low-copy transcripts by continuous SYBR® Green I monitoring during amplification. *Biotechniques*, **24**(6), 954–58.
14. Souazé, F, Ntodou-Thomé, A, Tran, CY, Rostene, W, Forgez, P. 1996. Quantitative RT-PCR: Limits and accuracy. *Biotechniques*, **21**, 280–85.
15. Wittwer, CT, Garling, DJ. 1991. Rapid cycle DNA amplification: time and temperature optimization. *Biotechniques*, **10**(1), 76–83.
16. Hartshorn, C, Anshelevich, A, Wangh, LJ. 2005. Rapid, single-tube method for quantitative preparation and analysis of RNA and DNA in samples as small as one cell. *BMC Biotechnol.*, **5**(1), 2.
17. Costafreda, MI, Bosch, A, Pinto, RM. 2006. Development, evaluation, and standardization of a real-time TaqMan Reverse Transcription-PCR assay for quantification of Hepatitis A virus in clinical and shellfish samples. *Appl. Environ. Microbiol.* **72**(6), 3846–55.
18. Abd-Elsalam, KA. 2003. Bioinformatic tools and guideline for PCR primer design. *Afr. J. Biotechnol.* 2(5), 91–5.
19. Freeman, WM, Walker, SJ, Vrana, KE. 1999. Quantitative RT-PCR: pitfalls and potential. *BioTechniques*. **26**, 112–25.
20. Giulietti, A, Overbergh, L, Valckx, D, Decallonne, B, Bouillon, R, Mathieu, C. 2001. An overview of real-time quantitative PCR, applications to quantify cytokine gene expression. *Methods*. **25**, 386–401.
21. Ginzinger, DG. 2002. Gene quantification using real-time quantitative PCR, An emerging technology hits the mainstream. *Exp Hematol.*, **30**, 503–12.
22. Hartshorn, C, Anshelevich, A, Wangh, LJ. 2005. Rapid, single-tube method for quantitative preparation and analysis of RNA and DNA in samples as small as one cell. *BMC Biotechnol.*, **5**, 1–13.
23. Heid, CA, Stevens, J., Livak, KJ, Williams, PM. 1996. Real time quantitative PCR. *Genome Res.*, 986–94.
24. Mackay, IM, Arden, KE, Nitsche, A. 2002. Real-time PCR in virology. *Nucleic Acids Res.*, **30**, 1292–305.
25. Mhlanga, MM, Malmberg, L. 2001. Using molecular beacons to detect single-nucleotide polymorphisms with real-time PCR. *Methods*, **25**(4), 463–71.
26. Pattyn, F, Speleman, F, De Paepe, A, Vandesompele, J. 2003. RTPrimerDB, the Real-Time PCR primer and probe database. *Nucleic Acids Res.*, **31**, 122–3.
27. Higuchi, R, Fockler, C, Dollinger, G, Watson, R. 1993. Kinetic PCR analysis, real-time monitoring of DNA amplification reactions. *Biotechnology*, **11**, 1026–30.
28. Tichopad, A, Dzidic, A, Pfaff, MW. 2002. Improving quantitative real-time RT-PCR reproducibility by boosting primer-linked amplification efficiency. *Biotechnol. Lett.*, **24**, 2053–56.
29. Niesters, HG. 2002. Clinical virology in real time. *J. Clin. Virol.*, **25** (3), 3–12.
30. Niesters, HG. 2004. Molecular and diagnostic clinical virology in real time. *Clin. Microbiol. Infect.*, **10**, 5–11.
31. Whitcombe, D, Theaker, J, Guy, SP, Brown, T, Little, S. 1999. Detection of PCR products using selfprobing amplicons and fluorescence. *Nat. Biotechnol.*, **17**, 804–07.

32. Wittwer, CT, et al. 1997. The LightcyclerTM a microvolume multisample fluorimeter with rapid temperature control. *Biotechniques*, **22**, 176–81.
33. Wittwer, CT, Herrmann, MG, Moss, AA, Rasmussen, RP. 1997. Continuous fluorescence monitoring of rapid cycle DNA amplification. *Biotechniques,* **22**, 130.
34. Holland, PM, Abramson, RD, Watson, R, Gelfand, DH. 1991. Detection of specific polymerase chain reaction product by utilizing the 5′–3′ exonuclease activity of *Thermus aquaticus* DNA polymerase . *Proc. Natl. Acad. Sci. USA*, **88**, 7276–80.
35. Tyagi, S, Kramer, FR. 1996. Molecular beacons-probes that fluoresce upon hybridization. *Nat. Biotechnol.*, **14**, 303–08.
36. Newton, CR, et al. 1989. Analysis of any point mutation in DNA. The amplification refractory mutation system (ARMS). *Nucleic Acids Res*., **17**, 2503–16.
37. Kiss, I, German, P, Sami, L, Antal, M, Farkas, T, Kardos, G, Kecskemeti, S, Dan, Á, Belak, S. 2006. Application of Real-Time RT-PCR utilising Lux (Light upon extension) Fluorogenic Primer for the rapid detection of avian influenza viruses. *Acta. Veterinaria Hungarica*, **54**(4), 525–33.
38. Täpp, I, Malmberg, L, Rennel, E, Wik, M, Syvänen, AC. 2000. Homogeneous scoring of single-nucleotide polymorphisms, comparison of the 5¢-Nuclease TaqMan assay and molecular beacon probes. *BioTechniques*, **28,** 732–8.
39. Zhou, L, Wang, L, Palais, R, Pryor, R, Wittwer, CT. 2005. High-resolution DNA melting analysis for simultaneous mutation scanning and genotyping in solution. *Clin. Chem*., **51** (10), 1770–7.
40. Palais RA., Liew MA., Wittwer CT. 2005. Quantitative heteroduplex analysis for single nucleotide polymorphism genotyping. *Anal. Biochem*., **346**(1) 167–75.
41. Lyon, E. 2001. Mutation detection using fluorescent hybridization probes and melting curve analysis.*Expert Rev. Mol. Diagn*., **1**(1), 92–101.
42. Bustin, SA, Nolan, T. 2004. Pitfalls of quantitative real-time reverse-transcription polymerase chain reaction. *J. Biomol. Tech*., **15**(3), 155–66.
43. Wong, ML, Medrano, JF. 2005. Real-time PCR for mRNA quantitation. *Biotechniques*, **39**(1), 75–85.
44. Reynisson, E, Josefsen, MH, Krause, M, Hoorfar, J. 2006. Evaluation of probe chemistries and platforms to improve the detection limit of real-time PCR . *J. Microbiol. Methods*, **66**(2), 206–16.
45. Mackay, IM. 2004. Real-time PCR in the microbiology laboratory. *Clin. Microbiol. Infect*., **10**(3), 190–212.
46. Yuan, JS, Reed, A, Chen, F, Stewart, CN. 2006. Statistical analysis of real-time PCR data. *BMC Bioinformatics*, **7**, 85.
47. Ririe, KM, Rasmussen, RP, Wittwer, CT. 1997. Product differentiation by analysis of DNA melting curves during the polymerase chain reaction. *Anal. Biochem*., **245**(2), 154–60.
48. Vet, JA, Marras, SA. 2005. Design and optimization of molecular beacon real-time polymerase chain reaction assays. *Methods Mol. Biol*., **288**, 273–90.
49. Vet, JA, Van der Rijt, BJ, Blom, HJ. 2002. Molecular beacons, colorful analysis of nucleic acids. *Expert Rev. Mol. Diagn*., **2** (1), 77–86.
50. Abravaya, K, Huff, J, Marshall R., et al. 2003. Molecular beacons as diagnostic tools, technology and applications. *Clin. Chem. Lab. Med*., **41** (4), 468–74.
51. Tan, W, Wang, K, Drake, TJ. 2004. Molecular beacons. *Curr. Opin. Chem. Biol*., **8**(5), 547–53.
52. Kubista, M, Andrade, JM, Bengtsson, M, Forootan, A, Jonak, J, Lind, K, Sindelka, R, Sjoback, R, Sjogreen, B, Strombom, L, Stahlberg, A, Zoric, N. 2006. The real-time polymerase chain reaction. *Mol. Aspects. Med*., **27**(2–3), 95–125.
53. Notomi, T, Okayama, H, Masubuchi, H, Yonekawa, T, Watanabe, K, Amino, N, Hase, T. 2000. Loop-mediated isothermal amplification of DNA. *Nucleic Acids Res*., **28**, 63.
54. Norihiro, T, Yasuyoshi, M, Hidetoshi, K, Tsugunori, N. 2008. Loop-mediated isothermal amplification (LAMP) of gene sequences and simple visual detection of products. *Nat. Protocols*, **3**, 877–82.

55. Hiyoshi, M, Hosoi, S. 1994. Assay of DNA denaturation by polymerase chain reaction-driven fluorescent label incorporation and fluorescence resonance energy transfer. *Anal. Biochem.*, **221** (2), 306–11.
56. Chen, X, Zehnbauer, B, Gnirke, A, Kwok, PY. 1997. Fluorescence energy transfer detection as a homogeneous DNA diagnostic method. *Proc. Natl. Acad. Sci., U S A.*, **94** (20), 10756–61.
57. Holland, PM, Abramson, RD, Watson, R, Gelfand, DH. 1991. Detection of specific polymerase chain reaction product by utilizing the 5′–3′ exonuclease activity of Thermus aquaticus DNA polymerase . *Proc. Natl. Acad. Sci. U S A.*, **88**(16), 7276–80.
58. Philip, SB, Richard SA, James, PK, Wittwer, CT. 1998. Homogeneous multiplex genotyping of hemochromatosis mutations with fluorescent hybridization probes. *Am. J. Pathol.*, **153**, 1055–61.
59. Ririe, KM, Rasmussen, RP, Wittwer CT. 1997. Product differentiation by analysis of DNA melting curves during the polymerase chain reaction. *Anal. Biochem.*, **245**(2), 154–60.
60. Pfaffl, MW. 2001. A new mathematical model for relative quantification in real-time RT-PCR. *Nucleic Acids Res.*, **29**(9), 45.

Chapter 4
New Trends in the Diagnosis and Molecular Epidemiology of Viral Diseases

Contents

E.A. Pestana et al., *Early, Rapid and Sensitive Veterinary Molecular Diagnostics - Real Time PCR Applications*, DOI 10.1007/978-90-481-3132-7_4,

4.1 Background

Despite intensive worldwide control programmes against infectious diseases, including vaccination programmes with the use of DIVA vaccines; mass culling (stamping out policies) and regulation of animal movements; various virus diseases still have a very high negative impact on animal health and welfare. The intensification of animal husbandry; centralisation of large groups of animals in industrial production units; globalization of trade in live animals and/or animal products, bedding and feeds; as well as increased tourism, are all considerable factors in the threat of devastating infectious diseases word-wide. The opening of borders between many countries such as in the European continent contributes greatly to the high-risk situation, where infectious agents may easily travel thousands of miles and then suddenly appear in areas where they are unexpected and probably even unknown. The sudden and unexpected appearance of any infectious disease in a new region, be it a country or a continent, may lead to a delayed or innaccurate diagnosis resulting in the uncontrolled spread of the disease agent to other susceptible populations of animals over large geographic areas. Recent major examples are incidences of foot-and-mouth disease (FMD) in the UK, the extension of rinderpest into the Somali plains and Rift Valley fever (RVF) spread into the Arabian Peninsula. The latest major problem is the occurrence, re-occurrence and rapid spread of influenza virus. All these exemplify the serious economic and social impact of the of highly contagious transboundary animal diseases (TADs).

4.1.1 Costs of Disease

The costs in dealing with TADs should be viewed both in terms of efforts to bring the disease under control and the consequent loss of livelihoods. As an example, the UK FMD outbreak in 2001 had a cost to the public sector estimated at around 4.5 billion Euros, and to the private sector at over 7.5 billion Euros. In considering such situations, the European Commission Scientific Committee on Animal Health and Animal Welfare report (adopted 24–25th April 2003) states, "Recent outbreaks of foot-and-mouth disease (FMD), classical swine fever (CSF) and avian influenza (AI) have occurred in several member states and resulted in the slaughter of large numbers of animals as well as severe economic consequences". The ethical problems arising as a consequence of current eradication strategies, as well as the social problems caused by the slaughter of large numbers of animals, must also be taken into account when considering the overall effects of TAD outbreaks.

4.1.2 Global Factors

Global climatic changes have had a direct measureable effect on the emergence and spread of viral diseases, in particular those transferred by insect vectors. This is clearly illustrated by the northward spread of bluetongue in Europe that reached Scandinavian countires by 2007. African horse sickness virus, a close relative to bluetongue virus in the Orbivirus genus of the Reoviridae family, has shown a similar pattern in spreading toward regions of the world where this disease was previously unknown. The main reason for this is that the insect vectors of the viruses (Culicoides) have extended their range since the climatic conditions have changed favourably. Other vector-borne viral diseases, such as African swine fever, show similar tendency of expansion in infected territories.

4.1.3 Other Diseases

In addition to TADs, there is a range of other viral diseases in our animal populations with a more restricted or endemic geographical character, but these still have a high economic and socio-economic impact. Despite intensive eradication programmes, these diseases still cause severe problems due to direct or indirect losses in the animal populations; increased treatment costs and decreased production rates. Diseases caused by viruses such as bovine herpesviruses, adenoviruses, bovine viral diarrhoea virus, bovine respiratory syncitial virus and bovine coronaviruses to name a few, are still commonly found in cattle populations all over the world and their eradication remains an important task. Pig populations in most of the world suffer from a high variety of viral diseases. Classical swine fever, African swine fever, Aujeszky's disease, foot-and-mouth disease, swine vesicular disease and other well-known and characterised viral diseases of swine are permanent targets of eradication programmes worldwide, with varying rates of success. For example, classical swine fever has been eliminated from the domestic pig herds of the EU, but wild boar populations are still infected in several countries, posing a high risk, though re-infection, for the whole continent. African swine fever is spreading, as previously indicated and there are no potent vaccines available. New variants of the virus have recently emerged in Africa, while in Sardinia remains infecected. Porcine respiratory and reproductive syndrome (PRRS) is a commonly occurring disease caused by an emerging arterivirus, first detected first in 1991 in the Netherlands, after which it rapidly spread all over the world. Other diseases are increasing in importance such as Postweaning multisystemic wasting syndrome (PMWS) or Porcine Dermatitis and Nephropathy Syndrome (PDNS), were the association with porcine cirvovirus type 2 was found, however, several questions are still unanswered in aetiology, pathogenesis and many other important factors of these diseases of swine.

4.1.4 Major Problems

Viral diseases of transboundary and/or endemic character are creating large problems in a high range of domesticated and wild animals worldwide. Due to the restricted space of this chapter, only a few examples are dealt with in detail in order to illustrate the role of viral diseases in the issue of animal health and welfare. In this context the emergence of "new" diseases is a very important issue, which needs special attention and focus. In addition to examples already given, a long list of emerging or re-emerging diseases in various host species, including humans could be drawn up; these include, diseases caused by Hantaviruses, Japanese Encephalitis Virus, HIV, Dengue Viruses, Menangle Virus, Australian Bat Lyssavirus, Ebola virus, Avian flu variants H5N1, SARS coronavirus, Nipah virus and Hendra Virus. Several of these diseases are of very serious global zoonotic concern.

4.1.5 Need to Improve Diagnosis

The above examples indicate that there is necessity to develop and use improved diagnostic methods to keep worldwide track of viral infections, both in human and animal populations. Powerful new methods have to be developed, which allow the prompt detection and identification of the viruses, supporting the animal health authorities and organizations for combating the viral diseases more effectively.

With regard to the global importance of viral diseases, including the emerging and re-emerging TADs, the international organisations such as the World Organisation for Animal Health (OIE, previously known as Office International des Epizooties), the Food and Agricultural Organisation (FAO) and the International Atomic Energy Agency (IAEA) are helping to combat the diseases at the international level. Simultaneously, the EU, the USDA and other international and national programmes and grants are supporting the work for the improved detection and control of TADs. As an example, in 1994 the FAO established an Emergency Prevention System (EMPRES) for Transboundary Animal and Plant Pests and Diseases in order to minimize the risk of such emergencies developing. In the "Animal Diseases" component of EMPRES, major TADs are targeted, including rinderpest and other epidemic animal diseases, such as contagious bovine pleuropneumonia, foot-and-mouth disease, contagious caprine pleuropneumonia, peste de petit ruminants, RVF, and lumpy skin disease. These are among the most contagious maladies and place a serious burden on the economies of the countries in which they occur.

4.1.6 Harmonization of Responses

In order to try and harmonise the efforts to combat TADs in a total of 172 Member Countries and Territories, the OIE organises work in the following structures of expertise, Specialist Commissions, Collaborating Centres, Reference Laboratories, OFFLU Working Groups and Ad hoc Groups. The OIE Collaborating Centres (OIE CCs) are centres of expertise in a specific designated sphere of competence relating to the management of general questions on animal health issues (for example epidemiology or risk analysis). In its designated field of competence, an OIE CC is providing expertise internationally. For details see www.oie.int, in OIE Mandate and Internal Rules for Collaborating Centres. Currently, the OIE has 24 CCs all over the world, dealing with many aspects of animal health, such as diagnosis, animal disease surveillance, risk analysis, epidemiology, food safety, animal welfare, vaccine evaluation, control of infectious diseases and veterinary medicinal products. A large part of work of the OIE CCs focuses on new and emerging diseases, and on national and international training (see: www.oie.int).

The National Veterinary Institute (SVA) in Uppsala was nominated and registered by the OIE as a CC in 2005. At present, our institute is termed, "OIE Collaborating Centre for the Biotechnology-based Diagnosis of Infectious Diseases in Veterinary Medicine". Our tasks involve biotechnology-based diagnosis; international standardisation and validation of the diagnostic methods and national and international training. These are summarized on the website of the CC: http://sva.se/oie-cc.

The previous research and development activities and molecular routine diagnostic activities of the CC have been summarised in review articles. More recent achievements in molecular diagnosis are summarized here; references are provided and strategies outlined as an updated report on the work of our OIE CC to further improve the conditions of animal health, welfare and food safety. This also serves as their basis for the use of the Standard Operation Procedures (SOPs), published in Chapter 5.

4.1.7 Application of Various PCR Methods in Routine Diagnostic Virology

4.1.7.1 Gel-Based and Real-Time PCR Assays as Novel Tools of Molecular Diagnosis

Shortly after the description of the PCR principle in 1985, our laboratory was one of the first to adapt the technique to diagnostic purposes and to develop routine diagnostic PCR assays. The first generation "*gel-based or classical*" PCR assays provided novel tools for the improved diagnosis of a high variety of TADs, as summarised in the reviews in the reference section.

4.1.7.2 Various Real-Time PCR Assays Are Further Improving the Diagnostic Facilities

From the middle of the 1990s a wide range of real-time PCR assays were developed and applied to the diagnosis of TADs and other infectious diseases, as well to the improved detection of pathogens in food and feed. Various real-time PCR assays including TaqMan, Molecular Beacons (MB), Primer-Probe Energy Transfer (PriProET), Scorpion Primers, dual probe systems such as those utilized in the LightCycler® (Roche), dye-labelled oligonucleotide ligation (DOL) and SYBR® Green; showed high sensitivity and specificity for the detection both of viral and bactertial pathogens.

Compared to the classical single or nested PCR methods, the new real-time PCR assays have a number of important advantages:

- Faster and higher throughput of tests.
- Real-time-PCR sensitivity close to or equal to traditional nested PCR even though a non-nested set-up is used.
- Amplified products detected by measuring fluorescence through the lid of the reaction vessel without having to open the system, thus minimizing the risk of contamination from laboratory environment, or carryover into nearby samples.
- Post-PCR handling of the products is not needed.
- The result of the PCR is not only either "positive" or "negative", but the real-time PCR assays allow a quantitative estimation of target nucleic acid in the sample.
- Real-time quantitative PCR is more accurate and less labour-intensive than current quantitative PCR methods.
- The hands-on time is greatly reduced, compared to traditional detection in agarose gels followed by ethidium bromide or GelRed™ staining.
- The principle of the real-time PCR allows automation of the procedure, and the use of a 96-well microtiter plate format, without the need for nested PCR, which makes it very practical to automate.
- Diagnosis can be further automated by using robots for DNA/RNA extractions and pipetting.
- Probes for real-time PCR can be labelled with a number of different fluorophores, which function as individual reporter dyes for different primer sets thus, real-time PCR is very suitable for the development of *multiplex PCR* systems
- Lower costs per detected agent, if the equipment can be used with a large enough number of samples.

Compared to the previous gel-based, classical, amplification assays, the real-time PCR techniques have a further advantages in that they allow the quantitative assessment of the targeted viral genomes. This can be very important in viral diagnosis, such as in the case of diagnosis of Postweaning Multisystemic Wasting Syndrome, where the number of viral particles or the viral loads of porcine circovirus type 2 (PCV2) have to be determined.

Currently, TaqMan and MB are the most commonly used real-time PCR methods in routine diagnostic laboratories consequently the SOPs in this book cover and mostly recommend these techniques. Simultaneously, other approaches, like the PriProET system, and LUX PCR are also under development in many laboratories and frequently applied. Recently other real-time PCR methods, such as the LATE PCR are also considered as robust, reliable assays for improved detection of various viruses, especially when adapted to simple tools, like portable PCR machines allowing on site diagnosis of the viral diseases, (see later).

It is very important to note that as diagnosticians we should not refer to "real-time PCR" in general, but should specify the method of real-time PCR precisely, as can be seen from the examples above. Thus the specification of the variant of real-time PCR, such as TaqMan or MB, is very important, considering that the various methods have different strengths and weaknesses, such as detection range, sensitivity, and specificity. It is also very important to understand that even within the same variant of the real-time PCR, significant differences can occur. As an example, various TaqMan methods developed for the detection of the same virus, can differ strongly in diagnostic specificity, sensitivity and other important parameters. Note the term diagnstic here, describing both sensitivity and specificity. Diagnostic refers to the performance of a test on "real" samples from the field. It is emphasized that the exact real-time PCR method used is clearly specified with very exact references to the methods applied, in order to avoid confusion and misunderstanding.

4.1.7.3 Importance of Determining the Diagnostic Sensitivity and Specificity of the Real-Time PCR Assays, as Well as the Target Detection Range

Some real-time PCR assays have been shown to detect as few as ten genome copies of the targeted viruses. This indicates very high analytical sensitivity. Concerning specificity, the majority of the assays are able to detect and amplify entirely the selected target nucleic acids and no cross-reactivity is disturbs the diagnosis. However, "wide spectrum", "pan-" or "general" PCR assays are also in use. Such methods are for example the "pan-pesti" PCR assays, which amplify very conservative regions of the pestivirus genomes (e.g., selected regions from the 5′NCR), and are able to amplify all tested pestiviruses, such as bovine viral diarrhoea virus (BVDV), classical swine fever virus (CSFV) and Border disease virus (BDV).

4.1.7.4 The Simultaneous Use of Various Real-Time PCR Methods, Allowing Wide-Range and More Specific Detection

Diagnostic work is effective and well organised when the laboratory is using an arsenal of wide-range ("general") real-time PCR assays for preliminary screening of the samples. Subsequently, the exact identification of the detected pathogens is made by "narrower range", highly specific PCR assays. Such assays allow not the exact identification of the virus variants(s), but allow the studies of molecular epidemiology.

4.1.8 Multiplex PCR in Routine Diagnosis

The multiplex PCR methods are based on the use of multiple primers to allow amplification of multiple templates within a single reaction, for example, analysis of a single nasal or rectal swab collected from an animal suffering from a respiratory disease, or from enteritis/diarrhoea syndrome, respectively. By performing multiplex PCR, we seek to diagnose all possible pathogens which can be considered to be causing the disease complex. In general the multiplex PCR assays are useful for diagnostic purposes, providing the diagnostician the ability to detect more than one infectious agent(s) in a single assay. The gel-based PCR assays allow the development of multiplex PCR; however, the real-time PCR is even more suitable for multiplexing. The reason is that the individual probes for the component assays can be labelled with different fluorophores, each of which functions as a specific colour reporter dye for one set of primers. Since the fluorescent probes emit at different colour wavelengths, it enables an easy multiplexing of the assays.

Compared to the single-target PCR techniques, the construction of multiplex assays can be rather complicated, considering the large number of primers required. The various primers might compete with each other, as they have to be placed in the same reaction mix of the classical nested PCR.

As stated above, the real-time PCR assays (using only single primer pairs) provide better possibilities for the construction of multiplex systems with multiple target components. Considering the diagnostic advantages of this principle, various multiplex PCR assays were developed at our laboratory, partly based on our own developments or in collaboration with EU project partner laboratories (http://www.multiplex-eu.org/ and http://www.labonsite.com/). For example, a multiplex (duplex) real-time PCR assay was developed and is now used in routine diagnosis for the simultaneous detection of bovine respiratory syncytial virus and bovine respiratory coronavirus, two pathogens important in the respiratory disease complexes of young calves. We have found that multiplex real-time PCR has the potential to produce considerable savings in time and effort, without compromising the robustness and sensitivity of the virus detection assays. However, as mentioned above, the competition of primers frequently hinder the development of potent multiplex PCR assays and the development trials may require long time and high costs, with uncertain success.

4.1.9 Simultaneous Detection of Viruses and the Complex Diagnosis, Development of "Multi PCR" Assays Simplify Diagnosis

Primer-competition might cause serious problems, not only in time and costs, but also in other aspects. For example, the levels of specificity and sensitivity of the PCR assay may strongly drop and in certain cases the system is simply not working at all in a multiplex arrangement. Our experience is that in such cases it is more practical not to force the co-amplification in the same reaction vessel, but rather to amplify the various viruses side-by-side on a microplate. This system is also multiplex, since the various viruses are simultaneously detected from a tested clinical sample. We term this approach "multi" PCR. By using automated systems, the multiplex and rapid detection of the various pathogens is achieved very rapidly in multi PCR. Various multi PCR assay are regularly used in our routine diagnostic section, providing reliable results. The use of such systems is highly recommended.

4.1.10 Robots are Accelerating Molecular Diagnosis and Provide Better Safety

Since the proper preparation of the targeted viral nucleic acids is a crucial step of the molecular diagnostic procedures, the introduction of nucleic acid extraction robots is a common practice today. The use of the robots is significantly accelerating the diagnostic procedures and provide more safety. Realizing the high need for these equipments, the industry is producing a wide range various types of the nucleic acid purifying robots. For example, the GenoVision M48 extraction robots (Biorobot M48 station, Qiagen, Norway), utilize magnetic separation of the target molecules. By comparing the results of nucleic acid preparations of the robot with manual procedures, we found the robot more efficient and reliable. This robot purified the nucleic acids simultaneously from 48 samples, within 2.5 h. The products were clean enough to be amplified directly in the PCR. In addition to high speed, robustness and low labour-input, a further advantage of the robots is the reduced risk of cross contamination between specimens. By the introduction of special tools laboratory practices and internal controls (mimics) it was possible to reduce the danger of false positivity and false negativity rather soon in the history of the diagnostic PCR.

The closed and automated systems of the robots provide strong safety for the PCR-based diagnostic assays. It is advisable to automate as many steps as possible in the diagnostic procedures. By simultaneously using nucleic acid extraction and pipetting robots with the real-time PCR machines, the laboratory can establish an automated diagnostic chain. Such chains have been established at our laboratory for the detection of several viruses. By the introduction of robots, high throughput and robust diagnostic assays have been established, with reduced manipulation requirement, less contamination risk and a very rapid diagnosis time, which is shortened from hours to minutes. Interestingly, the nucleic acid-based diagnosis will be similar to the ELISA-based diagnostic chains, using automated systems, which provide rapidity, robustness, low costs, reduced labour-requirement and an increased reliability of diagnosis.

4.1.11 Isothermal Amplification and the Use of Simple Thermo Blocks Can Replace Costly PCR Machines

Besides PCR, alternative methods of nucleic acid amplification are in progress. Such methods are for example the Invader or the LAMP technologies, which became recently common, providing nucleic acid amplification based diagnosis to less sophisticated and equipped laboratories. Without the need for costly PCR machines, the isothermal amplification methods use only simple thermo-blocks, which are affordable to moderately equipped laboratories and even to simple field laboratories. Thus, the isothermal amplification methods are optimal in poor countries and for bringing the diagnostic facilities closer to the outbreaks.

4.1.12 Portable PCR Machines

Portable PCR machines are constructed to bring the laboratory facilities closer to the field cases. Several companies are producing and optimising such machines today, which can easily be used under field conditions; run on batteries and that allow complete disinfection of the equipment, providing simple sample preparation and rapid results, without the need for specific training. For example, the machines of the Smiths Detection's Portable Veterinary Diagnostic Laboratory addresses the issue by avoiding any transportation of samples to a laboratory. With this system the laboratory is taken into the field. It has been designed with field veterinarians in mind, comprising a portable briefcase-sized PCR instrument and a disposable sample preparation unit. Together they provide rapid on-site identification, in a wide range of weather conditions, by veterinarians or other workers in animal health, who require no technical understanding of the PCR methodologies, just raise the questions about the suspicion of the occurrence of a viral disease.

Smiths Detection uses an advanced PCR chemistry called Linear After the Exponential PCR or LATE-PCR. This is an advanced form of asymmetric PCR that efficiently generates single-stranded amplicons under controlled conditions. LATE-PCR is further enhanced by additional technologies that improve sample preparation, suppress amplification errors, improve probe design for rapid high-resolution analysis of the amplified product, make multiplexing easier, allow for rapid DNA sequencing, and enhances data analysis. The combination of the simple, portable PCR machines and LATE PCR provides simple facilities for the on site diagnosis of viral diseases.

4.1.13 Studies of Molecular Epidemiology

Assays involving PCR yield specific DNA products that can be investigated and analysed by several means. The nucleic acid composition of the products can be determined through nucleotide sequencing. The obtained nucleic acid sequences can be analysed and compared with each other and with previously described sequences, obtained from large international databases, such as the GenBank. The rapid phylogenetic identification and tracing of viruses is termed "molecular epidemiology". Molecular epidemiological studies were conducted, for example, when genetic variants of classical swine fever virus (CSFV) were identified in several countries of Central Europe and when it was hypothesised that EU and US genotypes of the porcine respiratory and reproductive syndrome virus (PRRSV) evolved from a common ancestor, which is suspected to originate from Eastern Europe. Also, a molecular epidemiological approach was implemented as part of the Swedish BVD-control programme as a tool to facilitate the identification and tracing of routes of transmission of bovine viral diarrhoea virus (BVDV) between herds.

PCR amplification and comparative nucleotide sequence analysis allow not only the direct detection of the viruses but also retrospective genetic analysis of biological products and clinical samples. For example such approach can be applied to determine the identity of virus strains used for vaccine production. In one of our recent studies a virus "pick up" was observed in a commercially produced live attenuated BVDV vaccine. The results of this work emphasize that the contamination of commercially available live vaccines with exogenous virus strains (such as BVDV strain originating from foetal calf serum or from bovine cells) is a real risk factor in the bioindustry. Considering this risk, unequivocal analysis, including molecular methods, is needed to verify the authenticity of biological products, such as vaccines, foetal calf serum batches, cell lines, etc.

Molecular epidemiology is providing a considerable help to the animal health authorities, when combating various viral diseases. The occurrence of virus variants is detected; the spread of various variants is traced, allowing epidemiological analysis, cutting of ways of infection, implementation of prevention rules and other measures to control the spread of the diseases.

4.1.14 The OIE Rules for the International Standardization and Validation of the PCR-Based Diagnostic Assays

Considering the frequent occurrence of the viral diseases worldwide and the very intensive R&D development in the field of molecular diagnosis, there is a high need for the international standardization and validation of the developed assays. It is important that the veterinary diagnostic laboratories use identical, validated techniques, providing comparable results and allowing the same conclusions in different regions of the world.

This is the only way to combat TADs and endemic diseases effectively from a global aspects, following the "one world, one health" principle. National and international authorities require rigorous proof that the assays, used in various laboratories, are as reliable as possible and give identical results. International agencies like the OIE, the Joint FAO/IAEA Division, national research institutions and commercial companies make large efforts to agree on international standardization.

The OIE regularly publishes standards for the validation of diagnostic assays. Validation and international standardization of nucleic acid amplification-based diagnostic methods (like PCR) is the major task for the animal health authorities. The usual practice is that the specificity and the sensitivity of the newly developed PCR assays are compared to conventional assays, like virus isolation. The "in house" PCR assays will soon have to be replaced by validated and standardized procedures. The validation, standardization and quality control of PCR-based diagnostic techniques, which are now in progress, are a major task.

4.1.15 OIE Manual of Diagnostic Tests and Vaccines for Terrestrial Animals 2004, 2008

This Manual is providing a very important collection of diagnostic methods recommended and approved by the OIE, by the OIE Reference laboratories and Collaborating Centres. The Manual is available both in printed and in online forms, for the later see: http://www.oie.int/eng/normes/mmanual/A_summry.htm

Several chapters of the online version were updated in 2005 and 2006. Both versions will be completely updated and re-published in 2008.

4.1.16 Validation and Quality Control of Polymerase Chain Reaction Methods Used for the Diagnosis of Infectious Diseases (Chapter I.1.4. of the OIE Manual)

This chapter is available on: http://www.oie.int/eng/normes/mmanual/A_00014.htm

As co-author of Chapter I.1.4., I would like to quote several important parts of the text from the OIE Manual, "The purpose of this chapter is to extend the rules to a direct method of infectious agent detection, i.e. to adapt the principles of validation to the PCR assays. The experiences of the last decade indicate that the PCR techniques will eventually supersede many of the classical direct methods of infectious agent detection. It is clear that the PCR is replacing virus isolation or bacteria cultivation for the detection of agents that are difficult or impossible to culture. There are several reasons for this trend, including that virus isolation requires, i) the presence of replicating viruses; ii) expensive cell culture and maintenance facilities; iii) as long as several weeks to complete the diagnosis; and iv) special expertise, which is missing or diminishing today in many laboratories. Although PCR assays were initially expensive and cumbersome to use, they have now become relatively inexpensive, safe and user-friendly tools in diagnostic laboratories. The sensitivity and specificity of PCR is generally greater than isolation or capture ELISA procedures."

4.2 PCR Methods Used in Routine Molecular Diagnostics

4.2.1 OIE Collaborating Center for the Biotechnology-Based Diagnosis of Infectious Diseases in Veterinary Medicine

Our laboratory, in collaboration with other international partners has been actively involved in the validation processes, by following the stages of assay validation as suggested by the OIE. This is one of the reasons that the OIE granted our institute the title "OIE Collaborating Center for the Biotechnology-based Diagnosis of Infectious Diseases in Veterinary Medicine".

The tasks and mandate of the OIE CC are summarised in http://sva.se/oie-cc. One of the main tasks of the OIE CC is to contribute to the Manual of Diagnostic Tests and Vaccines for Terrestrial Animals, which is summarising the recommended diagnostic methods and procedures: http://www.oie.int/eng/publicat/en_standards.htm. A further task is to collaborate with international agencies, first of all with the Animal Production and Health Section, Joint FAO/IAEA Division, which is the Collaborating Centre of the OIE for ELISA and Molecular Techniques in Animal Disease Diagnosis. The two sister CCs of the OIE have joint efforts to conduct research in diagnostic development and in dissemination of results. By providing a view of the recent developments in molecular diagnostic virology and a collection of reliable SOPs, this book is a result if this collaboration.

4.2.2 Recent Developments in the Field of Diagnostic Virology at the OIE Collaborating Center for the Biotechnology based Diagnosis of Infectious Diseases in Veterinary Medicine

This book is providing a comprehensive collection of SOPs, developed for the improved diagnosis of viral diseases and recommended to diagnostic laboratories. To complete the range of the proposed methods, to show further possibilities and to provide a wider view on the recent developments, the following paragraphs are summarising several very recent directions of diagnostic development in molecular diagnosis of viral diseases. Some of these methods are already used in the diagnostic laboratories, the other open new trends for the diagnosis in the near future.

4.2.2.1 Improved Detection of Foot-and-Mouth Disease Virus (FMDV) with a Novel, Robust Real-Time PCR

Since the simultaneous detection of various serotypes of the virus is an important task in the field, the "Multiplex PCR" EU project developed a novel, robust quantitative real-time PCR assay for the simultaneous detection of all the seven serotypes of FMDV. This method is based on the *Primer-Probe Energy Transfer (*PriProET) principle and is targeting the 3D gene of the virus. The assay was validated for the efficacy to detect all the seven known FMDV serotypes. The test method was linear over a range of at least seven orders of magnitude and the detection limit was below the equivalent of 10 genomic copies of the virus. Analysing recent African probang samples the method was able to detect FMDV in materials from both cattle and buffalo. When compared to traditional virus cultivation, the virus detection sensitivity was similar but the PriProET method can provide a laboratory result much faster than virus cultivation. Thus, a complex diagnosis of a foot-and-mouth-disease outbreak can be accomplished within several hours, providing a powerful new tool for the animal health authorities.

4.2.2.2 Solid Phase Microarrays in Veterinary Diagnostic Virology, Based on Padlock Probes

A solid phase microarray system was developed for the simultaneous detection of *foot-and-mouth disease, vesicular stomatitis and swine vesicular disease viruses*, using padlock probes. The application of padlock probes for detection of pathogens is a very recent trend in molecular diagnosis. Padlock probes are circularizable oligonucleotides useful for highly multiplex genetic studies. These probes have the capacity to detect simultaneously thousands of different target sequences in a single multiplex array system. Each viral nucleic acid serves as template for a padlock probe equipped with a unique sequence (tag) associated to that specific target. Circularized probes are amplified with a single universal primer pair and the fluorescently labelled products are then sorted, using the tag sequences, on a microarray. The assay principle is straightforward comprising a few internally controlled reaction steps in a single vessel. Run-times were comparable to real-time PCR, but with the benefit that the presence of several viruses and their various serotypes can be analyzed within the same reaction. Although the cost for oligonucleotides and array slides is higher than that for conventional PCR assays, it is reduced in proportion to the number of assays made. In this format, multiplex detection using padlock probes and microarrays could have implications in more effective screening for viruses causing similar vesicular symptoms and in turn facilitate rapid counteractions, especially in case of FMD outbreaks.

4.2.2.3 Padlock Probes for Broad-Range Detection and Subtyping of Avian Influenza Viruses

Using padlock-probe chemistry for multiplexed preamplification and microarray for detection, we developed an assay for the simultaneous detection and subtyping of avian influenza viruses (AIV). The assay has the outstanding feature to identify both the hemagglutinin (HA) and the neuraminidase (NA) surface antigens of AIV from a single reaction. We tested 77 influenza strains, representing the entire assortment of HA and NA, and 100% (77/77) of the samples were identified as AIV and 97% (75/77) were correctly subtyped. The specificity of the assay was determined testing heterologous pathogens. The results indicate that the assay is a useful and robust tool for high throughput rapid detection and typing of AIVs, with advantages compared to conventional methods.

In summary, the padlock probes, adapted to microarray formats, provide novel means of powerful and very complex novel molecular diagnosis. Compared to real-time PCR assays, the padlock probe based microarrays are inferior in sensitivity, but they allow a very multiplex diagnosis and the simultaneous analysis of thousands of specimens in the same system.

4.2.2.4 Novel TaqMan® and Primer-Probe Energy Transfer Assays for the Improved, Universal Detection of Hepatitis E Virus

Hepatitis E virus (HEV) is an important cause of food- and waterborne diseases in countries with poor sanitation, but recently it is getting more frequent also in regions of the world where the health services are of high standard. Previously the disease cases were observed in connection to travelling, recently zoonotic transmission is also suspected, i.e., a direct route of infection, from animals to humans. For the improved detection of the virus, two real-time PCR methods were developed and compared, a TaqMan® and Primer-Probe Energy Transfer (PriProET) assay. These robust, highly sensitive methods provide valuable diagnostic tools to investigate zoonotic transmission, to detect the virus in the food chain. They are used in research related to the potential of hepatitis E virus to cross the species barrier. By using the two novel PCR assays a broad range of viruses were detected, representing all the four genotypes of HEV. On comparison, the TaqMan® assay showed higher fluorescence values for positive samples. On the other hand, the PriProET better tolerated the point mutations in the target nucleic acids. Thus, the PriProET provides a more powerful tool to detect new variants of HEV. The two real-time PCR assays are useful novel tools for virus detection and for molecular epidemiology. In addition, the assays provide novel tools to study the biology of the viruses, including the transmission between various species and the zoonotic aspects of HEV infections.

4.2.2.5 Development of a Real-Time PCR Assay Based on Primer-Probe Energy Transfer for the Detection of Swine Vesicular Disease Virus

Based on primer-probe energy transfer (PriProET), we developed a real-time PCR assay to detect swine vesicular disease virus (SVDV). The assay was highly sensitive with a detection limit corresponding to five copies of viral genome equivalents, and had a high specificity demonstrated by testing of heterologous viruses. A major advantage of the PriProET chemistry is tolerance toward mutations in the probe region. Melting curve analysis directly after PCR, with determination of probe melting point, confirmed specific hybridisation of the SVDV strains. Eight of twenty SVDV strains tested, revealed shifted melting points that indicated mutations in the probe region, which were confirmed by nucleotide sequencing. With the PriProET system there is a chance to identify phylogenetically divergent strains of SVDV, which may appear negative in other probe-based real-time PCR assays. Moreover, any difference in melting points may provide an indication of divergence in the probe region. The described SVDV PriProET assay, with high sensitivity, specificity, and tolerance toward mutations in the probe region, provides a powerful tool for the improved and rapid detection of SVDV. Furthermore, it allows a reduced turnaround time and the use of high-throughput, automated technology.

4.2.2.6 Simple and Rapid Detection of Swine Vesicular Disease Virus with a One-Step Reverse Transcriptase Loop-Mediated Isothermal Amplification Assay

A one-step reverse transcriptase loop-mediated isothermal amplification (RT-LAMP) assay was developed recently for the improved detection of swine vesicular disease virus (SVDV). The assay provided a wide SVDV detection range, since all the 28 tested isolates of this virus were tested positive. Simultaneously, it yielded high specificity; because all tested heterologous viruses gave negative results, such as foot-and-mouth disease virus (FMDV) and vesicular stomatitis virus (VSV).

Since SVDV, FMDV and VSV cause very similar symptoms, the highly specific detection and identification of SVDV is very important. By testing RNA from clinical samples including nasal swabs, serum and faeces, the performance of the RT-LAMP was compared to a real-time PCR assay. When testing nasal swabs and serum, the sensitivity of the assays was approximately equivalent. Interestingly, by testing faecal samples the RT-LAMP assay performed better. According to our hypothesis, inhibitory substances probably less influenced the RT-LAMP assay and this could be the reason of the better performance.

The RT-LAMP assay has several strong features, which prove the applicability as a powerful new tool of SVDV detection, these include:

(i) As an isothermal amplification method, this does not require costly PCR machines, just a simple thermoblock.
(ii) Rapidity, since results are obtained within 30–60 min.
(iii) It can be highly specific and sensitive, as described above
(iv) Test reading is simple since the results are visualised either by gel- electrophoresis or by the naked eye through the addition of SYBR® Green (Fig. 4.1)

a) b)

Fig. 4.1 Simple detection of swine vesicular disease virus (SVDV) by loop-mediated isothermal amplification assay (LAMP) assay. By applying isothermal amplification, no expensive PCR machines are needed, only a simple thermo-block. The reading of the results is also simple by running a gel-electrophoresis (**a**) or by naked eye through the addition of SYBR® Green, in order to show the positive results by green colour (**b**). Three positive samples are seen on the left and one negative on the right, both on panels "a" and on "b". In addition, "a" shows a size marker on the very left. Picture b) kindly provided by AL Blomström

Since the RT-LAMP can easily be performed in modestly equipped field laboratories and it can be adapted to mobile diagnostic units, it provides novel tool for the "front line" diagnosis of swine vesicular disease, an important TAD. Considering the above-listed advantages of the RT-LAMP technology, we are adapting this method to the improved diagnosis of a range of viral diseases in various host species, including cattle.

4.2.2.7 Subtyping and Pathotyping of Avian Influenza Viruses with a One-Step Real-Time SYBR® Green RT-PCR Assay

For the rapid subtyping and pathotyping of avian influenza viruses (AIV) a one-step real-time SYBR® Green RT-PCR assay was developed. Primers were selected to target highly conserved nucleotide stretches that flank the cleavage site of the haemagglutinin (HA) gene of AIVs. By sequencing the amplified PCR products, both the subtype and in case of H5 subtypes even the pathotype of the detected AIV can rapidly be identified. By testing 27 strains of AIV and nine heterologous pathogens, including influenza B and C, and various avian viruses, the specificity of the assay was confirmed. Since the subtype and pathotype determination were completed within approximately 6 h, the SYBR® Green RT-PCR assay provides a powerful new too in the arsenal of influenza diagnostics.

4.2.3 Ultra Rapid Nucleic Acid Amplification and Nucleotide Sequencing Analysis

Considering that during the TAD outbreaks one of the most important requirements is the prompt diagnosis, a one-step real-time PCR assay was developed at our OIE CC. The assay is based on the amplification of genomic sequences from the HA gene, for the rapid and simultaneous detection of a broad spectrum of influenza viruses, including highly pathogenic avian influenza variants. Several prototypes of real-time PCR systems, which use the superconvection principle ("Superconvection QPCR"; Alpha Helix, Uppsala, Sweden), were used both for amplification and for cycle sequencing reactions (Figure 4.3). Identification and differentiation of pathogenicity variants of AIV (HPAI or LPAI) were determined, based on sequencing of HA gene and the results were obtained in less than 2 h. The procedures included 30 min for RNA purification, 30 min for superconvection QPCR, cycle 20 min for sequencing reactions with superconvection and 20 min for sequencing by capillary electrophoresis (up to 700 bases). The wide screening of different subtypes of avian influenza in a single QPCR, followed by rapid sequencing that covers cleavage site of HA gene, allow monitoring the viral load of influenza strains in wild birds and in farmed poultry. Furthermore, the method could provide a very rapid and highly reliable molecular diagnosis in a possible pandemic influenza A scenario. Further optimization is in progress.

4.2.4 *Proximity Ligation, Novel Means of Protein Detection by Nucleic Acid Amplification*

Proximity ligation is based on the following principle: The recently established proximity ligation mechanism enables sensitive high-capacity protein detection, identification and measurement by converting the detection of specific proteins to the analysis of DNA sequences. Proximity ligation enables a specific and quantitative transformation of proteins present in a sample into nucleic acid sequences. As pairs of so-called proximity probes bind the individual target protein molecules at distinct sites, these reagents are brought in close proximity. The probes consist of a protein specific binding part coupled to an oligonucleotide with either a free 3′- or 5′-end capable of hybridizing to a common connector oligonucleotide. When the probes are in proximity, promoted by target binding, then the DNA strands can be joined by enzymatic ligation. The nucleic acid sequence that is formed can then be amplified and quantitatively detected in a real-time monitored polymerase chain reaction. This convenient assay is simple to perform and allows highly sensitive protein detection.

Our partner laboratories (OLINK and SVANOVA, Uppsala) reported that detection sensitivities similar to those of nucleic acid-based detection reactions were achieved for the rapid detection of foot-and-mouth disease virus.

Compared to AgELISA, the sensitivity of proximity ligation proved to be higher in several approaches. Thus, proximity ligation of proteins can be of great value for early diagnosis of infectious disease and in biodefense.

At present, we are adapting the proximity ligation technique to the detection of surface antigens of various viruses, such as avian influenza virus.The combination of nucleic acid and antigen detection approaches will hopefully yield a more complex, multilateral diagnosis of TADs. In addition, replacement of Antigen-ELISA with more sensitive assay would be a great step in the diagnosis of infectious diseases.

4.2.5 *A Simple Magnetic Bead-Based Microarray for Detection and Discrimination of Pestiviruses*

A novel assay was developed for the rapid detection and discrimination of pestiviruses, i.e., BVDV types 1 and 2, CSFV and BDV, by using magnetic bead detection of PCR products on microarrays. After amplification, the PCR products are hybridized onto an array, followed by visualization with streptavidin-coated magnetic beads. The simple set-up allows visualization of results on the array either with the naked eye or a microscope, and makes this novel assay suitable for use in a modestly equipped laboratory.

A panel of pestiviruses comprising members of all the four accepted species was used to evaluate the assay. Other post-PCR detection methods (e.g., gel electrophoresis and suspension microarray) were used as comparisons for the determination of the detection sensitivity of the assay. The results clearly indicate that the assay provides a novel, robust and highly sensitive and specific method for the improved detection and discrimination of viral pathogens. Considering the simplicity of the assay, and the very simple detection procedure in particular, this magnetic bead-based assay offers a powerful and novel technology for molecular diagnostics in virology.

4.2.6 Detection of an Emerging Pestivirus in Cattle and Further Characterization by Means of Molecular Diagnostics and Reverse Genetics

During a study on Bovine Viral Diarrhoea (BVD) epidemiology in Thailand, by using indirect antibody ELISA, an antigen ELISA and PCR, a pestivirus was detected in heat-inactivated serum sample of a calf.

The PCR products were sequenced and the comparative nucleotide sequence analysis showed that this virus was closely related to a recently described atypical pestivirus (D32/00_'HoBi') that was first isolated from a batch of foetal calf serum collected in Brazil (Fig. 4.2).

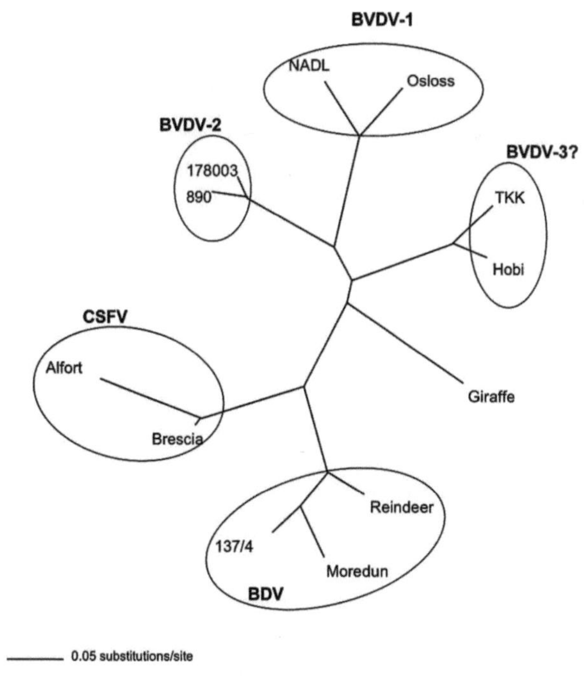

Fig. 4.2 Molecular epidemiology, identification of new pestiviruses, including TKK. Unrooted phylogram generated from a fragment of the 5´NCR sequences of selected representatives of each known species within the genus *Pestivirus*, including the tentative pestivirus of giraffe, the D32/00_'HoBi' and Th/04_KhonKaen (TKK). The sequences of pestivirus reference strains and previously described strains were obtained from the GenBank. The phylogram shows the phylogenetic position of the newly detected pestiviruses Hobi and TKK as new variants of BVDV (BVDV-3?), see more in Liu et al. [26]

It was also demonstrated that the Thailand virus (called Th/04_KhonKaen, or TKK) was circulating in the herd. The study was the first to report a natural infection in cattle with a virus from this group of atypical pestiviruses.

The data suggested that these viruses might be spread in cattle populations in various regions of the world. If so, these atypical bovine pestiviruses can have important implications for BVD control and for the biosafety of vaccines and other biological products, produced with foetal calf serum.

To study these important issues, we needed a "live" virus, which can be replicated and studies from various aspects. Since the serum was inactivated, virus isolation was not possible. Thus, we applied methods of reverse genetics, in order to "reconstruct" the inactivated virus from the inactivated serum sample.

By using transfection of the viral nucleic acids, the virus was reconstructed and re-gained its capacity to grow in cell cultures. Full-genome characterization of the new virus was performed after recuperation of the inactivated virus through transfection.

This characterization, and phylogenetic analysis based on the full genome sequence, demonstrated that the virus was closely related to BVDV, suggesting that TKK and other HoBi-like pestiviruses constitute a third genotype of BVDV, i.e. BVDV type 3 (BVDV-3).

4.2.7 Molecular Epidemiology, New Approaches

Recently we can see a considerable progress not only in the development of various nucleic acid detection methods, but also in molecular epidemiology. Traditionally, inferring molecular phylogeny used to be performed by distance-based method, i.e. neighbour joining for single genes.

In a recent study at our OIE CC, a Bayesian approach was exploited to analyze five genetic regions of BVDV genome (5-UTR, N^{pro}, E2a, E2b, and NS3) for 68 taxa retrieved from GenBank. The best performance of the method was achieved when analyzing a genetic region with appropriate proportions of conserved and variable sites or a combined dataset composed of all five genetic regions. In the future, Bayesian method combined with other traditional tree-building methods can be used to estimate a more reliable viral phylogenetic tree and to study the emerging and/or occurrence new variants of BVDV.

4.2.8 Further Trends, New Directions in Molecular Diagnostic Virology

The listed examples provide a short view about recent developments in molecular diagnostic virology. This field is developing rapidly and many novel approaches of virus detection and characterization are improving the means of virus detection and characterisation. There is no space to list many approaches, just several examples.

4.2.8.1 Full-Genome Amplification of Viral Genomes

To systematically identify and analyze the viral genomes, e.g., the15 HA and 9 NA subtypes of influenza A virus, we need reliable, simple methods that not only characterize partial sequences but analyze the entire influenza A genome.

By the selection and construction of specific sets of primers, it is possible to generate full-length cDNAs, to subtype viruses, to sequence their DNA, and to construct expression plasmids for reverse genetics systems.

4.2.8.2 Full-Genome Sequencing of Viral Genomes

By the use of new sequencing machines, such as the 454 sequencers (Roche, http://www.454.com/), 30–60 million nucleotides can be sequenced within several hours. The technique is extremely important in the genetic research, including the investigation and comparison of the viral genomes. By this way very exact information can be obtained about the full-length sequences of the tested viral genomes. These long sequences provide more reliable data for determination of evolutionary aspects, relationships of viruses, viral subpopulations and many other important aspects of molecular virology (Fig. 4.3).

In order to facilitate the detection of such viruses a range of molecular methods have been developed, in order to genetically characterize new viruses without prior *in vitro* replication or the use of virus-specific reagents. In the recent *metagenomic*

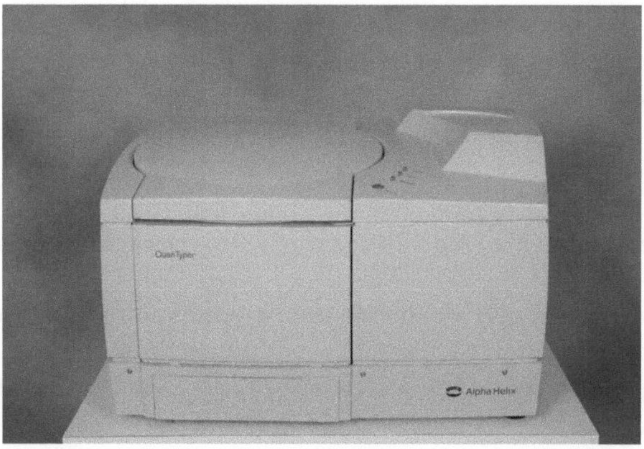

Fig. 4.3 Rapid sequencing of the viral genomes. QuanTyper^{TM}-48, from AlphaHelix Molecular Diagnostics AB, is the first real-time thermo cycling instrument utilizing SuperConvection^{TM} to speed up the cycling process. This is achieved by subjecting the samples to high g-force, (via centrifugation), while cycling the temperature. The result is an increased mixing of the sample fluid that improves both the temperature homogenization and the kinetics of the reaction. This makes the QuanTyper-48 very fast but also very flexible with a vast sample volume range, accommodating reactions from 20 to 200 μL in volume using standard PCR-tubes. Typically a 20 μL real-time PCR assay is ready within 15 min while a 200 μL reaction takes up to 50 min to complete. Since such viruses can also be important factors in various diseases, it is important to work on methods, which facilitate their detection

studies viral particles were detected in uncultured environmental and clinical samples, by using random amplification of their nucleic acids, prior to subcloning and sequencing. Already known and novel viruses were then identified by comparing their translated sequence to those of viral proteins in public sequence databases.

A wide range of specimens was tested by viral metagenomic approaches, such faeces, serum, plasma, respiratory secretions and organ suspensions, seawater, near shore sediments, etc. Selection of samples with high viral loads, purification of viral particles, removal of cellular nucleic acids, efficient sequence-independent amplification of viral RNA and DNA, recognizable sequence similarities to known viral sequences and deep sampling of the nucleic acid populations through large scale sequencing can all improve the possibilities to detect new viruses and to increase the yield. Viral metagenomic approaches provide novel opportunities to generate an unbiased characterization of the viral populations in various organisms and environments. For example, such techniques led to the detection of new parvoviruses, termed bocaviruses, in lower respiratory tract infections of children.

4.2.9 Viral Metagenomics, Search for Unknown Viruses

Characterization of hitherto unknown viruses is often hindered due to reasons such as,

- Inability or very poor capacity to grow in cell cultures
- Low copy number or unusual virion structure, hindering the detection and identification of the virions with electron microscopy
- Being a minor part of mixed viral infections, where the co-infecting other viruses are dominant
- Limited antigenic/serological cross-reactivity, hindering the detection by antigen-antibody assays, such as AgELISA, immunofluorescence or immunohistochemistry
- The lack of nucleic acid identity or similarity to known viral sequences, thus, remaining undetected in nucleic acid hybridization assays
- Remaining undetected even by PCR assays, due to unique genome structures, etc.

By detecting a range of new viruses recently, viral metagenomics have broadened the range of known viral diversity and opened a very interesting new area in virological research and diagnosis. It is worth to mention, that the above-mentioned approaches are not completely different and separated for each other, because they are frequently used together, in a complex assay. For example, full length amplification and full-length sequencing of the viral genomes is a logical combination.

Here only very few examples were given for the new trends. This area is extremely rapidly expanding in the recent days and a wide range of techniques is under development to conduct research in viral metagenomics and to detect unknown viruses.

4.2.10 Summary and Recommendations

This chapter gives a short summary from the *OIE Collaborating Center for the Biotechnology-based Diagnosis of Infectious Diseases in Veterinary Medicine*, summarising briefly several of the recent developments in molecular diagnostic virology. The purpose is to provide a view and a basis for the better understanding of the SOPs, provided in the subsequent chapters of this book.

Various aspects of diagnosis of viral diseases are shortly discussed are brief recommendations are given. Technical facilities are outlined and selected references are provided. This guidance is helping the reader to see the background of the SOPs and to follow the technical suggestions, in order to diagnose the viral diseases rapidly and effectively, following the rules of international standardisation and validation of the diagnostic assays.

Bibliography

1. Ackermann, M, Engels, M. 2006. Pro and contra IBR-eradication. *Vet. Microbiol.*, **113**, 293–302. Review.
2. Agüero, M, Fernández, J, Romero, LJ, Zamora, J, Sánchez, C, Belák, S, Arias, M, Sánchez–Vizcaíno, JM. 2004. A highly sensitive and specific gel-based multiplex PCR assay for the simultaneous and differential diagnosis of African swine fever and classical swine fever in clinical samples. *Vet. Res.*, **35**, 551–563.
3. Allander, T, Tammi, MT, Eriksson, M, Bjerkner, A, Tiveljung-Lindell, A, Andersson, B. 2005. Cloning of a human parvovirus by molecular screening of respiratory tract samples. *Proc. Natl. Acad. Sci. USA.*, **102**, 12891–6. Erratum in *Proc. Natl. Acad. Sci.,USA*, 102, 15712.
4. Bálint, A, Baule, C, Pálfi, V, Belák, S. 2005. Retrospective genome analysis of a live vaccine strain of bovine viral diarrhoea virus. *Vet. Res.*, **36**, 89–99.
5. Banér, J, Gyarmati, P, Yacoub, A, Hakhverdyan, M, Stenberg, J, Ericson, O, Nilsson, M Landegren, U, Belák, S. 2007. Microarray-based molecular detection of Foot-and-mouth disease, Vesicular stomatitis and Swine vesicular disease viruses, using padlock probes. *J. Virol. Methods*, **143**, 200–06.
6. Belák, S. 2005. The molecular diagnosis of porcine viral diseases, a review. *Acta. Vet. Hung.*, **53**, 113–24.
7. Belák, S. 2007a. Experiences of an OIE Collaborating Centre in molecular diagnosis of transboundary animal diseases, a review. *Dev. Biol.* (Basel), **128**, 103–12.
8. Belák, S. 2007b. Molecular diagnosis of viral diseases, present trends and future aspects a view from the OIE Collaborating Centre for the Application of Polymerase Chain Reaction Methods for Diagnosis of Viral Diseases in Veterinary Medicine. *Vaccine*, **25**, 5444–52.
9. Belák, S, Hakhverdyan, M. 2006. Recent achievements and trends in the molecular diagnosis of bovine viral diseases – a view from the "OIE Collaborating Centre for the Application of Polymerase Chain Reaction Methods for the Diagnosis of Viral Diseases in Veterinary Medicine". *Dtsch. Tierärztl Wochenschr*, **113**, 129–33.
10. Belák, S, Thorén, P. 2001. Molecular diagnosis of animal diseases, some experiences over the past decade. *Expert Rev. Mol. Diagn.*, **1**, 434–43.

11. Belák, S. 2004. Validation and quality control of polymerase chain methods used for the diagnosis of infectious diseases. Office International des Epizooties (OIE), Manual of Diagnostic Tests and Vaccines for Terrestrial Animals (mammals, birds and bees). Fifth Edition, Chapter I.1.4., 30–36.
12. Blomström, AL, Hakhverdyan, M, Reid, SM, Dukes, JP, King, DP, Belák, S, Berg, M. 2008. A one-step reverse transcriptase loop-mediated isothermal amplification assay for simple and rapid detection of swine vesicular disease virus. *J. Virol. Methods*, **147**, 188–93.
13. Chan, CH., Lin KL., Chan, Y, Wang, YL, Chi, YT, Tu, HL, Shieh, HK, Liu, WT. 2006. Amplification of the entire genome of influenza A virus H1N1 and H3N2 subtypes by reverse-transcription polymerase chain reaction. *J. Virol. Methods*, **136**, 38–43.
14. Crowther, JR, Unger, H, Viljoen, GJ. 2006. Aspects of kit validation for tests used for the diagnosis and surveillance of livestock diseases, producer and end-user responsibilities. *Rev. Sci. Tech.*, **25**, 913–35. Review.
15. Delwart, EL. 2007. Viral metagenomics. *Rev. Med. Virol.*, **17**, 115–31.
16. Farkas, T, Antal M, Sámi L, Germán P, Kecskeméti S, Kardos G, Belák S, Kiss, I. 2007. Rapid and simultaneous detection of avian influenza and Newcastle disease viruses by duplex polymerase chain reaction assay. *Zoonoses Public Health*, **54**, 38–43.
17. Fernández, J, Agüero, M, Romero, L, Sánchez, C, Belák, S, Arias, ML, Sánchez-Vizcaíno, JM. 2008. Rapid and differential diagnosis of foot-and-mouth disease (FMD), swine vesicular disease (SVD) and vesicular stomatitis (VS) by a new multiplex RT-PCR assay, analysing clinical samples. *J. Virol. Methods*, **147**, 301–11.
18. Gyarmati, P, Mohammed, N, Norder, H, Blomberg, J, Belák, S, Widén, F. 2007. Universal detection of hepatitis E virus by two real-time PCR assays, TaqMan and Primer-Probe Energy Transfer. *J. Virol. Methods*, **146**, 226–35.
19. Gyarmati, P, Conze, T, Zohari, S, LeBlanc, N, Nilsson, M, Landegren, U, Banér, J, Belák, S. 2008. Simultaneous genotyping of all hemagglutinin and neuraminidase subtypes of avian influenza viruses using padlock probes. *J. Clin. Microbiol.*, **46**, 1747–51.
20. Hakhverdyan, M, Rasmussen, TB, Thorén, P, Uttenthal, Å, Belák, S. 2006. Development of a real-time PCR assay based on primer-probe energy transfer for the detection of swine vesicular disease virus. *Arch. Virol.*, **151**, 2365–76.
21. Hjertner, B, Meehan, B, McKillen, J, McNeilly, F, Belák, S. 2005. Development of invader squared assay for the detection of African swine fever DNA and comparison to PCR based assays. *J. Virol. Methods*, **124**, 1–10.
22. Hoffmann, E, Stech, J, Guan, Y, Webster, RG, Perez, DR. 2001. Universal primer set for the full-length amplification of all influenza A viruses. *Arch.Virol.*, **146**, 2275–89.
23. Kiss, I, Germán, P, Sámi, L, Antal M, Farkas, T, Kardos, G, Kecskeméti, S, Dán, Á, Belák, S. 2006. Application of LUX (light upon extension) flourogenic primer utilizing real-time RT-PCR for the rapid detection of avian influenza viruses. *Acta. Vet. Hung.*, **54**, 525–33
24. Larsen, LE. 2000. Bovine respiratory syncytial virus (BRSV), a review. *Acta. Vet. Scand.*, **41**, 1–24.
25. LeBlanc, N, Gantelius, J, Schwenk, J, Ståhl, K, Blomberg, J, Andersson-Svahn, H, Belák, S. 2008. Development of a magnetic bead microarray for the simultaneous and simple detection of four pestiviruses. *J. Virol. Methods*, in press.
26. Liu, L, Kampa, J, Belák, S, Baule, C. 2008. Identification of a new species of pestivirus as genotype-3 Bovine Viral Diarrhoea Virus (BVDV-3) based on full-length genome sequence analysis. Submitted
27. McKillen, J, Hjertner, B, Millar, A, McNeilly, F, Belák, S, Adair, B, Allan, G. 2006. Molecular beacon real-time PCR detection of swine viruses. *J. Virol. Methods*, **140**, 155–65.
28. Moennig, V, Houe, H, Lindberg, A. 2005. BVD control in Europe, current status and perspectives. *Ani. Health Res. Rev.*, **6**, 63–74.
29. Nordengrahn, A, Gustafsdottir, SM, Ebert, K, Reid, SM, King, DP, Ferris, NP, Brocchi, E, Grazioli, S, Landegren, U, Merza, M. 2008. Evaluation of a novel proximity ligation assay for the sensitive and rapid detection of foot-and-mouth disease virus. *Vet Microbiol.*, **127**, 227–36.

30. Rasmussen, T, Uttenthal, Å, de Stricker, K, Belák, S, Storgaard, T. 2003. Development of a novel quantitative real-time PCR assay for the simultaneous detection of all serotypes of foot-and-mouth disease virus. *Arch. Virol.*, **148**, 2005–21.
31. Sámi, L, Ursu, K, McKillen, J, Kecskeméti, S, Belák, S, Kiss, I. 2007. Simultaneous detection of three porcine viruses by multiplex PCR. *Acta. Vet. Hung.*, **55**, 267–76.
32. Sanchez, JA, Pierce, KE, Rice, JE, Wangh, LJ. 2004. Linear-After-The-Exponential (LATE)-PCR, An advanced method of asymmetric PCR and its uses in quantitative real-time analysis. *Proc. Natl. Acad. Sci. USA.*, **101**, 1933–38.
33. Segalés, J, Calsamiglia, M, Olvera, A, Sibila, M, Badiella, L, Domingo, M. 2005. Quantification of porcine circovirus type 2 (PCV2) DNA in serum and tonsillar, nasal, tracheo-bronchial, urinary and faecal swabs of pigs with and without postweaning multisystemic wasting syndrome (PMWS). *Vet. Microbiol.*, **111**, 223–9.
34. Spatz, SJ., Rue, CA. 2008. Sequence determination of a mildly virulent strain (CU-2) of Gallid herpesvirus type 2 using 454 pyrosequencing.*Virus Genes*. **36**, 479–89.
35. Stadejek, T, Stankevicius, A, Storgaard, T, Oleksiewicz, MB, Belák, S, Drew, TW, Pejsak, Z. 2002. Identification of radically different variants of porcine reproductive and respiratory syndrome virus in Eastern Europe, towards a common ancestor for European and American viruses. *J. Gen. Virol.*, **83**, 1861–73.
36. Ståhl, K, Kampa, J, Alenius, S, Persson Wadman, A, Baule, C, Aiumlamai, S, Belák, S. 2007. Natural infection of cattle with an atypical 'HoBi'-like pestivirus – implications for BVD control and for the safety of biological products. *Vet. Res.*, **38**, 517–23.
37. Ståhl, K, Kampa, J, Baule, C, Isaksson, M, Moreno-Lopez, J, Belák, S, Alenius, S, Lindberg, A. 2005. Molecular epidemiology of bovine viral diarrhoea during the final phase of the Swedish BVD-eradication programme. *Prev. Vet. Med.*, **72**, 103–8; discussion 215–9.
38. Vilcek, S, Belák, S. 1998. Classical swine fever virus, discrimination between vaccine strains and European field viruses by restriction endonuclease cleavage of PCR amplicons. *Acta. Vet. Scand.*, **39**, 395–400.
39. Xia, H, Liu, L, Wahlberg, N, Baule, C, Belák, S. 2007. Molecular phylogenetic analysis of bovine viral diarrhoea virus, a Bayesian approach. *Virus Res.*, **130**, 53–62.
40. Yacoub, A., Kiss, I, Zohari, S, Hakhverdyan, M, Czifra, G, Mohamed, N, Gyarmati, P, Blomberg, J, Belák, S. 2009. A simple assay for rapid subtyping and pathotyping of avian influenza viruses. *In preparation*.

Chapter 5
Disease Diagnosis Using Real-Time PCR Specific Procedures for Important Veterinary Pathogens

This chapter provides SOPs for the following processes and diseases,

1. Detection of Avian Influenza A matrix gene by real time TaqMan RT-P CR.
2. H7 Eurasian Real Time PCRs for the detection and pathotyping of Eurasian H7 avian influenza isolates.
3. One Step RT PCR for detection of H5 & H7 avian influenza & cleavage site sequencing.
4. Eurasian H5 avian influenza Real Time PCR.
5. Detection of Rift Valley fever virus by Real-Time Reverse Transcription -PCR.
6. Swine vesicular disease virus one-step RT-LAMP.
7. Detection of African swine fever virus DNA using the isothermal ASFV vp72 655T Invader squared assay.
8. Real-time PCR detection and quantification of porcine viruses using molecular beacons
9. Swine vesicular disease virus PriProET two-step real-time PCR.
10. Slope/end point analysis of Invader data.
11. African Horse Sickness TaqMan RT-PCR.
12. Bluetongue SYBR-Green RT-PCR.
13. BTV Serotype 4 SYBR® GREEN RT-PCR.
14. CSFV Specific Real Time RT-PCR.
15. Realtime RT PCR detection of Influenza virus matrix gene Realtime RT PCR detection of velogenic Newcastle Disease fusion protein
16. Preparation of Silica particles for nucleic acid extraction.
17. Boom-silica RNA extraction (GuSCN, phenol, Silica).
18. Mab based competitive ELISAs for H5, and H7 antibody detection in avian sera.
19. Type A, H5, and H7 Avian Influenza antigen detection ELISAs.
20. Ribonucleic acid extraction from samples using TRIzol Reagent.

E.A. Pestana et al., *Early, Rapid and Sensitive Veterinary Molecular Diagnostics - Real Time PCR Applications*, DOI 10.1007/978-90-481-3132-7_5,

Published by Springer Science+Business Media B.V., Dordrecht 2010.

21. Ambion Magnetic Beads Extraction (96-well).
22. Svanodip® FMDV-Ag penside test.
23. FMDV PLA assay.
24. Procedure for using the Molecular Diagnostics Suite.
25. One step TaqMan® RT-PCR for diagnosis of FMDV and related vesicular viruses.
26. Operation of the Stratagene Mx4000/Mx3005P for real-time PCR. One-step RT-PCR amplification of RNA from vesicular disease viruses.
27. Differentiation of sheep and goat poxviruses by real time PCR

Introduction

Standard Operating Procedures

The aim of this book is to show developments in PCR, particularly those involving real time and real time on-line applications. This Chapter deals with a variety of PCR systems in order to illustrate such developments. The evolution of methods involves many stages and applied analyses and inevitably different stages are reflected in publication of operating procedures or protocols. This is true in the following descriptions which, although called Standard Operating Procedures (SOP), reflect the aforementioned variation in validation stages reached to fully merit the term standard.

The key to useful operating procedures is to define the "fitness for purpose" of a test in whatever format. It seems obvious, but the exact purpose of a test has to be clearly defined before an SOP is relevant to anyone seeking, or adopting, such a method. Often the SOP describes a method that is employed as a routine test procedure, working well, but without vigorous and proper validation. Often a test is without a defined purpose or target application and often a test is devised firstly as a research tool and then adapted for routine or diagnostic use. These "un-validated" SOPs are often the "best" procedures, or the only ones we have, but they should be regarded as useful guidelines which will require proper validation before adapted as definitive SOPs. It is to be expected that, in the case of a defined purpose, these laboratory protocols will undergo the robust validation sooner than later.

It must be strongly emphasised that there are few systems involving real time and real time on-line PCR that are validated to a sufficient degree to be acceptable as routine tests (i.e. the procedure can be repeated in your lab without much "tweaking" to produce the same level of result confidence) and that can be reproduced from an SOP in an international range of laboratories. In the main, this situation will stay with us in the short term, although there are international moves to review methods which have similar fitness for purpose and this should lead to better "standards" and harmonisation at least, where needed. The situation should improve since the usage of real time PCR is becoming common-place and more methods are being applied which will require at least comparative assessments (i.e. internal and external quality assurance and ring testing) of diagnostic sensitivity and specificity.

It should be expected (and is on the whole true) that reference laboratories provide the "best" SOPs since they have a mandate to develop tests fit for purpose and they have enough material (samples) and experience to allow validation principles to be fulfilled to eventually to define systems that are proven to work under a set of defined criteria. For diagnostic purposes this could involve a portfolio of several "near to validated", or "validated", set of laboratory procedures to address a specific purpose. In this book we target foot-and-mouth disease as example. Here the vital areas of the test are the sample frame, sample taking and handling (transport, storage), the receiving, the extraction, the testing, the reporting and the actions taken. In short we have a defined fitness for purpose involving all situations from field to laboratory reporting. The PCR testing is only one component of the SOP. The greater complexity of describing needs is illustrated in the chapter.

The SOPs here reflect a good cross section of what is mentioned above. In fact SOPs covering more than the real time PCR are included since they are used in tandem with PCR methods to provide a test system which better fits the fitness for purpose needs. This illustrates that SOPs should be geared to fitness for purpose and that there may be two or more tests adding up to fulfil the criterion.

The SOPs published are as given by the various contributors and reflect different styles. There is no internationally accepted format for showing SOPs and this is another area where international organisations can help standardise and improve the quality of SOP writing.

Readers with further interest in SOPs quoted should always contact the named person for further information and clarification. It should also be noted that often there is a great deal more information available from the donor of the SOP than is shown or indicated in the SOP and this is often the validation data held internally. The publication of validation data is a difficult area and although the OIE has a mechanism for test registration there is a need for an information exchange for the routine updateable dissemination of methods and data.

The SOPs have kindly been provided from various sources. There are differences in layout; font and style most of which have been preserved in this book. In practice, as already stated, there is no accepted standard for SOP presentation.

The reader will get protocols for tests from other sources and will be subjected to different degrees of detail and explanation. It is worth repeating that wherever possible a contact person should be ascribed to answer questions on the protocol given and who is also responsible for updating or altering the protocol.

SOP 1. Detection of Avian Influenza A Matrix Gene by Real Time TaqMan RT-PCR

SOP VI 493 Edition 2 18/09/07

This protocol is the standard operating procedure used by the avian influenza CRL at the Veterinary Laboratories Agency.

If you have any technical queries please contact aiwrl@vla.defra.gsi.gov.uk

1 Introduction

1.1 Purpose/Scope of This Protocol

To rapidly detect influenza RNA extracted from clinical and chick embryo amplified samples and provide an early warning for the presence of influenza virus in a sample.

1.2 Background Information

Real time PCR is a method that has been introduced relatively recently. The technology combines DNA amplification with detection of the products in a single tube. In the case of influenza A viruses an extra reverse transcriptase step is required to convert the RNA into cDNA. This format is highly beneficial as it removes the significant contamination risk caused by opening tubes for post-PCR manipulation. It is also less time consuming than gel based analysis and can supply a quantitative result. Current detection methods are based on changes in fluorescence proportional to the increase in product. Fluorescence is monitored during each PCR cycle to provide an amplification plot, allowing the user to follow the reaction.

The PCR is based on detection of a conserved region of the influenza A virus matrix gene using fluorogenic hydrolysis type probes and will detect all influenza A subtypes.

2 Materials

2.1 Chemicals and Reagents

2.1.1 Primers and Probe

Sep 1	AGA TGA GTC TTC TAA CCG AGG TCG
Sep2	TGC AAA AAC ATC TTC AAG TCT CTG
SePRO	FAM-TCA GGC CCC CTC AAA GCC GA-TAMRA

2.1.2 Real Time PCR Master Mix

This utilises the Qiagen Onestep RT-PCR kit (Cat No. 210212). Reagents not contained in the kit can be obtained from other suppliers. Volumes indicated below will be sufficient for 10 × 25 μL reactions.

DEPC treated water	Ambion or similar	137.5 μL
(x5) Qiagen 1 step RT PCR buffer		50 μL
Rox ref dye (pre diluted 1:500 in DEPC water)	Stratagene or similar	3.75 μL
Qiagen dNTP mix		10 μL
Sep1(50 μM)	Sigma or similar	2 μL
Sep2(50 μM)	Sigma or similar	2 μL
SePRO(30 μM)	Sigma or similar	2.5 μL
25 mM magnesium chloride	Promega or similar	12.5 μL
RNAsin (40 U/μL)	Promega or similar	1 μL
Qiagen 1 step RT PCR enzyme mix		10 μL

2.1.3 Real Time PCR M Gene Standards

These are quantified RNA standards extracted from infected allantoic fluid of known titre.

2.2 Equipment

- Microcentrifuge tubes (1.5 mL)
- –18°C or lower freezer
- –70°C or lower freezer
- Vortex mixer
- Pipettes
- Microcentrifuge (with rotor for 2 mL tubes)
- Sterile, Rnase-free pipette tips with aerosol barrier

- Mx3000/4000 real time RT-PCR quantitative machine
- Mx3000/4000 real time RT-PCR plates/strips
- Mx3000/4000 real time RT-PCR plate caps

3 Procedure/Method

3.1 Preparation of PCR Master Mix and Loading

3.1.1 Preparation of PCR master mix and loading of real time plate/strip(s) to be carried out in the PCR clean room.
3.1.2 Make up master mix sufficient for the number of samples to be tested.
3.1.3 Thoroughly mix the master mix and centrifuge for 30 s to remove bubbles.
3.1.4 Aliquot 23 μL of master mix per well of the real time plate/strip(s).
3.1.5 Loosely place the plate caps on the plate.
All wells will need to be covered even though only a portion of the plate may be used.
3.1.6 Bring the plate/strip(s) out of the clean room and place on ice before addition of sample RNA and RNA standards.

3.2 Preparation of RNA Standards

3.2.1 Extract RNA using appropriate method
3.2.2 Carefully prepare dilutions of RNA in water in the range 10^0, 10^1, 10^2, 10^3, 10^4. Mix each dilution by vortexing and briefly centrifuge. Make sufficient of each dilution so that a number of aliquots of each dilution will be made. Aliquot in volumes that the RNA does not have more than 3 freeze/thaws. Store aliquots at –70°C.
3.2.3 Each standard to be run as indicated on the worksheet

3.3 Manual Addition of Samples and Standards

3.3.1 Referring to plate layout, add 2 μL of the sample RNA to the 23 μL of master mix.
Take care to avoid cross contamination of samples at this stage.
Change gloves frequently and do not hold tips containing RNA above the incorrect wells.

3.3.2 Add 2 μL of real time PCR standards to appropriate wells based on your worksheet layout. Also include a no template control (NTC) well as a negative control, using 2 μL water as your template. Re-apply caps to the rest of the plate/strip(s).

3.3.3 Once RNA is added, fit caps to all wells.
It is important that the caps are fitted firmly and correctly onto the wells before being used on the real time machine.

3.3.4 If the real time plate/strip(s) are not to be loaded onto the real time machine straight away, keep the plate/strip(s) on ice until ready to test.

3.4 Reverse Transcriptase and PCR

3.4.1 Place the real time plate/strip(s) in the appropriate real time machine.

3.4.2 Incubate the reactions with the following thermocycling profile

RT step	50°C for 30 min
	95°C for 15 min
PCR step (x 40 cycles)	95°C for 10 s
	60°C for 20 s

Collect fluorescence data during the 60°C annealing step using the ROX and FAM filters and the endpoint read option with 2 reads if using the Mx4000. The Mx3000 will only allow one reading per 20 s.

3.4.3 Note that the lamp on both real time machines requires 20 min to warm up. Open a new file on the machine and use the real time quantitative PCR(multiple standards) option for the experiment type. The thermocycle profile and plate setup can be re-entered for each experiment or imported from previous experiments if desired.

3.4.4 When the plate setup and thermoprofile windows have been entered, select run to start thermocycling. A storage window will automatically open. Save the file in an appropriate folder on the hard drive of the computer.

3.4.5 The complete run takes approximately 1 h and 50 min. If the machine is not being used by anyone else after you, select the option to turn the lamp off at the end of the run as the lamps have a limited lifespan.

4 Results

4.1 Analysis and Display of Results

4.1.1 The fluorescence data can be viewed during, and after, the PCR reaction using the raw data plots tab in the run section.

4.1.2 When the run is complete make a record of the file name and its storage location and save a backup of the file in a separate location.

4.1.3 To analyse the data select the Analysis section button and select the wells to be examined in the Analysis/Setup window.

4.1.4 To view the results click on the Results tab in the Analysis section and view the amplification plots.

There are four options for analysing the fluorescence:

- *R* (Multicomponent view) displays the *raw fluorescence* in arbitrary units.
- *dR* displays the *baseline-corrected fluorescence.*

  As all reactions and wells will start with a slightly different fluorescence reading this option sets a baseline value of 0 to all plots. This correction is determined by the fluorescence values obtained during the initial rounds of the PCR. The adaptive baseline algorithm calculates the best baseline for each plot individually.

- *Rn* displays the *fluorescence normalised* to the passive reference dye (ROX).
  This allows for fluctuations in fluorescence, which are not due to cleavage of the Taqman probe.

- *dRn* displays the baseline-corrected normalised fluorescence.
  The dRn option with the ROX channel switched on is the most appropriate option for analysis of the data.

4.2 Interpretation of Results

4.2.1 Analyse the data by comparing the results obtained for the positive standards and NTC.

4.2.2 Construct a standard curve with the positive standards. As a guideline the curve should have the following criteria:

Efficiency >80%
Slope between –3.1 and –3.8
R2 value >0.980

4.2.3 Check the Ct values for any robot extraction controls.
These should reproducibly amplify with a Ct value ±2 Ct's either side of their predetermined Ct value of 30.Greater deviations from the predetermined Ct value suggest that,

- The M gene positive control may have degraded (if Ct greater than predetermined value).

- Software settings on the instrument are inaccurate/inappropriate and the fluorescence cut-off level excessively high/low and is affecting the Ct value of the predetermined M gene positive control. It may be possible to reset the software after the experiment to restore the expected Ct value provided the other controls are in order.
- Deviation (±2 Cts) may also occur where software settings on the instrument are inaccurate/inappropriate and the fluorescence cut-off level excessively high/low and is affecting the Ct value of the predetermined M gene positive control. It may be possible to reset the software after the experiment to restore the expected Ct value provided the other controls are in order.

4.2.4 If all controls are within acceptable limits analyse the data for the test samples.

4.2.5 An increase in fluorescence will be observed at an early stage for positive samples. The NTC and negative samples should not result in an increase in fluorescence above the baseline.

4.2.6 Criteria for assessing whether a sample is positive or negative are as follows,

- Ct value <30 – samples are M gene positive.
- Ct value 30–35 with sigmoidal/logarithmic appearance – samples are weak positive.
- Ct value >35 samples are negative if the amplification plot has a linear shape. These plots may represent spurious probe degradation or non-specific fluorescence.

4.2.7 All positive results are repeated and confirmed by use of virus isolation or conventional gel based PCR and sequencing before reporting in order to establish the subtype of influenza A virus detected.

5 Bibliography

1. Chang-Won L, Suarez, DL. 2004. Application of real time RT-PCR for the quantification and competitive replication study of H5 and H7 subtype avian influenza virus. *J. Virol. Methods*, **119**, 151–8.
2. Spackman, E, Senne, DA, Myers, TJ, Bulaga, LL, Garber, LP, Perdue, ML, Lohman, K, Daum, LT, Suarez, DL. 2002. Development of a real time reverse transcriptase PCR assay for type A influenza virus and the avian H5 and H7 haemagglutination subtypes. *J. Clin. Microbiol.*, **40**, 3256–60.

6 Appendix 1

Testing New Batches of Reagents

New batches of probe and primers should be tested when they are first re-suspended. Probe and primers should be aliquoted once re-suspended to avoid repeated freeze thawing.

Test new reagents in the same real time PCR as the current reagents using the positive standards and NTC according to this protocol. If the new reagents are fit for purpose both sets of reagents should give equivalent results.

SOP 2. H7 Eurasian Real Time PCRs for the Detection and Pathotyping of Eurasian H7 Avian Influenza Isolates

SOP VI.536 Edition 4 11/04/07

This protocol is the standard operating procedure used by the avian influenza CRL at the Veterinary Laboratories Agency.

If you have any technical queries please contact aiwrl@vla.defra.gsi.gov.uk

1 Introduction

1.1 Purpose/Scope of This Protocol

This protocol provides the details required to (i) detect Eurasian isolates of H7 avian influenza virus (AIV) by RealTime PCR and also to (ii) pathotype any H7 positive specimen by amplicon sequencing.

1.2 Background Information

1.2.1 Two H7 RealTime PCRs are presented in this document. Both are rapid, sensitive and specific method for the detection of Eurasian H7 AIV isolates. In brief, the two H7 RealTime PCR methods differ as follows,

1.2.2 "HA2" H7 RealTime PCR, Validation studies at VLA have shown this to be the more sensitive of the two H7 detection methods, amplifying a product within the HA2 region of the H7 gene. Detection utilises a H7-specific hydrolysis ("TaqMan") probe with fluorescence emission in RealTime.

1.2.3 Cleavage Site ("CS") H7 RealTime PCR, While validation studies have shown this to be the less-sensitive of the two H7 RealTime PCRs, this method amplifies a product which spans the H7 gene cleavage site (CS) region. Sequencing of this H7 CS amplicon allows for successfulpathotyping of the H7 AIV as either low pathogenic (LP) or highly pathogenic avian influenza (HPAI).

1.2.4 Validation at VLA has revealed both H7 RealTime PCR approaches to be effective in detecting H7 AI directly from clinical specimens, ie there is no need to first isolate the suspect H7 isolate in embryonated eggs by growth to a high titre. H7 RealTime PCR has been described previously [1], but here primer and probe design was tailored for the detection of N American H7 AI isolates.

1.2.5 For the methods described in this protocol, haemaglutinnin (HA) gene sequences from Eurasian H7 isolates (mainly 1995–2005) were aligned. This served to optimise primer and probe design for the detection of Eurasian H7 AIV isolates. The *HA2* H7 RealTime PCR was designed at VLA, while the *CS* H7 RealTime PCR was designed by Mr John Voermans and Dr Guus Koch at CIDC-Lelystad in the Netherlands.

1.2.6 In this protocol there is considerable commonality for both the HA2 and CS H7 RealTime PCRs as regards Materials (Section 3), Procedure/Method (Section 4) and Results (Section 5). However, it is important to note the differences between these two H7 RealTime PCRs as regards (i) Chemicals and Reagents (3.2) for primers and probes, and (ii) the results interpretation guide for the two methods, i.e. Sections 5.2.5 and 5.2.6.

1.2.7 In addition, the CS H7 RealTime PCR provides the additional opportunity to pathotype by amplicon sequencing as stated in Procedure/Method (Section 4.6). Examination of this H7 CS amplicon sequence will determine whether the H7 specimen is low pathogenic (LP) or highly pathogenic avian influenza (HPAI)

2 Safety

It is your laboratory's responsibility to ensure all work described in this protocol is conducted to a high safety standard. This includes an awareness of risks relating to eg dangerous or toxic chemicals, potentially hazardous procedures etc. Here local safety rules in your laboratory should be understood by all relevant members of staff.

3 Materials

3.1 Documentation and Software

Worksheet for RealTime plate layout.

Instruction manual for RealTime PCR platform and software in use in your laboratory.

3.2 Chemicals and Reagents

3.2.1 "HA2" H7 RealTime PCR Primers and Probe

LH6H7	5′-GGC CAG TAT TAG AAA CAA CAC CTA TGA-3′
RH4H7	5′-GCC CCG AAG CTA AAC CAA AGT AT-3′
H7pro11	5′FAM-CCG CTG CTT AGT TTG ACT GGG TCA ATC T-BHQ-3′
H7R	5′-GAC CTT CCC ATC CAT TTT CA-3′
H7-TM1	5′-FAM-AAC CCG CTA TGG CAC CAA ATA GGC CTC-BHQ-3′

3.2.2 HA2 H7 RealTime PCR Master Mix

This utilises the Qiagen OneStep RT-PCR kit (Cat No. 210212).

Reagents not contained in the kit can be obtained from other suppliers.

Volumes indicated below will be sufficient for 10 × 25 μL reactions, i.e., divide the Master Mix into 10 × 23 μL volumes and add 2 μL extracted RNA to each.

DEPC treated water (i.e. RNAse-free)	Ambion or similar	139 μL
(x5) Qiagen 1 step RT PCR buffer		50 μL
Rox ref dye (pre-diluted 1:500 from Stratagene stock in DEPC water)	Stratagene or similar	3.75 μL
Qiagen dNTP mix		10 μL
LH6H7 (50 μM)	Sigma or similar	2 μL
RH4H7 (50 μM)	Sigma or similar	2 μL
H7pro11 (30 μM)	Sigma or similar	1.25 μL
25mM magnesium chloride	Promega or similar	12.5 μL
RNAsin (40 U/μL)	Promega or similar	1 μL
Qiagen 1 stepRT PCR enzyme mix		10 μL

3.2.3 CS H7 RealTime PCR Master Mix

This utilises the Qiagen OneStep RT-PCR kit for the HA2 H7 RealTime PCR.

The only differences are the choice of primers and probe.

The reagents below will be sufficient for 10 × 25 μL reactions, i.e., divide the Master Mix into 10 × 23 μL volumes and add 2 μL extracted RNA to each.

DEPC treated water (ie RNAse-free)	Ambion or similar	139 μL
(x5) Qiagen 1 step RT PCR buffer		50 μL
Rox ref dye (pre-diluted 1:500 from Stratagene stock in DEPC water)	Stratagene or similar	3.75 μL
Qiagen dNTP mix	Sigma or similar	10 μL
H7F (50 μM)	Sigma or similar	2 μL
H7R (50 μM)	Sigma or similar	2 μL
H7-TM1 (30 μM)	Sigma or similar	1.25 μL
25mM magnesium chloride	Promega or similar	12.5 μL
RNAsin (40U/μL)	Promega or similar	1 μL
Qiagen 1 stepRT PCR enzyme mix		10 μL

3.2.4 H7 Controls

- Negatives
  Include at least two "no template controls" (NTCs) where 2μL RNase free water are added.

- H7 extraction control (positive)
  E.g., an aliquot of inactivated H7 virus from your repository. This should serve as an RNA extraction control (manual or robotic). The Ct value of this extracted RNA should be predetermined earlier according to your decision, but a "low positive" titre is advisable eg a Ct value of approximately 30. This is because a low titre extraction control is much more likely to indicate problems with RNA extraction, whereas a high titre (ie lower Ct) extraction may not indicate this. Higher titre extraction controls may also risk contaminating neighbouring reaction wells. Ideally use in duplicate for each H7 RealTime experiment. Enter as "unknown" (or equivalent) during experimental set-up using the software on the RealTime instrument.

- H7 RNA dilution series ("standards")
  Use any other Eurasian H7 RNA preparation from extracted egg fluid as ×10 fold dilution series (at least 4 dilutions). Ideally this should be a duplicate dilution series. This series (entered appropriately as "standards" with relative quantities indicated by using the software while setting up the RealTime experiment) can be used to determine the H7 RealTime PCR efficiency. It is important to include low positive H7 controls which correspond to approximately 10 EID_{50}/mL and 100 EID_{50}/mL of H7 AIV in the sample prior to RNA extraction – this will relate to defining the H7 RealTime PCR positive/negative cutoff, where the corresponding Ct value will differ in the HA2 and CS versions (Sections 5.2.5 and 5.2.6, below).

- Quantification of H7 positive controls
  The H7 positive control material may have been quantified by using classical virus titre determination in embryonating chickens' eggs or by using

M-gene RealTime in quantitative mode with a series of standards which had themselves been obtained from AIV of known titre.

3.3 Equipment

- Microcentrifuge tubes (1.5 mL)
- Correctly functioning RNA extraction robot
- –70°C or lower freezer
- 20 μL robot barrier tips
- Pipettes
- –18°C or lower freezer
- Sterile, Rnase-free pipette tips with aerosol barrier
- Vortex mixer
- Appropriate plasticware for the RealTime PCR platform of your choice
- Microcentrifuge (with rotor for 2 mL tubes)
- RealTime PCR platform of your choice, where you are satisfied that both the instrument are in correct working order.

4 Procedure/Method

4.1 Preparation of PCR Master Mix and Loading

4.1.1 Preparation of PCR master mix and loading of RealTime plasticware for reactions to be carried out in the PCR clean room.
4.1.2 Make up master mix sufficient for the number of samples to be tested.
4.1.3 Thoroughly mix the master mix and centrifuge for 30 s to remove bubbles.
4.1.4 Aliquot 23 μL of master mix per well of the RealTime plate/strip(s). NB: Ensure the plasticware (ie plates/strips and covers/caps) is appropriate for the chosen RealTime PCR platform.
4.1.5 Use plasticware to loosely cover the plate/strips accordingly
4.1.6 Bring the plate/strip(s) out of the PCR clean room and place on ice before addition of sample RNA and controls.

4.2 Preparation of H7 RNA Controls

4.2.1 H7 positive and negative controls are as described above in Section 3.2.5.

4.2.2 Extract H7 RNA manually or robotically according to the manufacturer's instructions.

4.2.3 Include H7 RNA positive controls "standards" (ideally in duplicate) as a dilution series in each H7 RealTime PCR experiment (Section 3.2.5, Paragraph 3).

4.3 Manual Addition of Samples and Standards

4.3.1 Referring to plate layout, add 2 μL of the sample RNA to the 23 μL of master mix. Where possible, use a dedicated multi-channel pipette to load multiple samples e.g., eight samples per strip.
NB: Take care to avoid cross contamination of samples at this stage. Change gloves frequently and do *not* hold tips containing RNA above the incorrect wells. It is advised to cover strips loosely which have not yet been filled with caps (strip covers) to further minimise the risk of cross contamination.

4.3.2 Add 2 μL of RealTime PCR controls to appropriate wells based on the worksheet layout. Re-apply covers to the rest of the plate/strip(s).

4.3.3 Once RNA is added, fit caps (strip covers) to all wells.

4.3.4 It is important that the caps or covers are fitted firmly and correctly onto the wells before insertion into the RealTime instrument.

4.3.5 If the RealTime plate/ strip(s) are not to be loaded into the instrument immediately, keep the plate/strip(s) on ice until ready to test.

4.4 Robotic Addition of Samples

4.4.1 Gloves to be worn at all times.

4.4.2 Work according to the manufacturer's instructions in order to extract RNA from specimens.

4.4.3 At VLA we prefer to add extracted RNA to the RealTime plate manually by using a multichannel pipette. Once addition of RNA is complete, re-seal the plate containing the eluted RNA and store at –70°C.

4.4.4 Manually add 2 μL of RealTime PCR controls (positive and negative) to appropriate wells as described in 4.3.2.

4.4.5 Ensure that the caps/strip covers are completely flat across the whole plate.

4.4.6 The plate is now ready to be run on a RealTime PCR instrument.

4.4.7 If the plate is not to be loaded into the RealTime instrument immediately, keep plate on ice until required.

4.5 Reverse Transcription (RT) and PCR

4.5.1 Place the RealTime plate/strip(s) in the appropriate RealTime instrument.

4.5.2 The thermocycling profile is identical for *both* the *HA2* and *CS* H7 RealTime PCRs.

These are also identical to that used for the H5 RealTime PCR which enables more than one of these assays to be ran simultaneously in one plate

RT step 50°C for 30 min
95°C for 15 min

PCR step (×40 cycles) 95°C for 10 s
54°C for 30 s
72°C for 10 s

NB: Collect fluorescence data at end of 54°C step using the ROX and FAM filters.

4.5.3 Open a new file using the instrument software to define all the necessary RealTime PCR experimental parameters. This is conducted according to the manufacturer's software instructions.

4.5.4 Ensure that the software files are distinguished ie *HA2* or *CS* H7 RealTime PCR. If running both H7 RealTime PCRs and/or the H5 RealTime PCR (4.5.2, above) in the same experiment, then name the file appropriately.

4.5.5 The complete run takes approximately 2 h, but this may vary according tothe make of instrument. If the machine is not booked for immediate use by another colleague upon completion of the run, select the option (if available) to turn off the lamp at the end of the run. This is because certain makes of lamp have a limited lifespan.

4.6 Molecular Pathotyping by Sequencing the CS H7 RealTime PCR Amplicon

4.6.1 Here the amplicon generated by the *CS* H7 RealTime PCR can be directly purified from the reaction mix after completion of this RealTime PCR. There is *no* need to electrophorese this H7 amplicon through an agarose gel prior to purification. Amplicon purification and sequencing is as described by protocol "OneStep RT PCR for detection of H5 & H7 avian influenza and cleavage site sequencing", with the difference that primers H7F and H7R (3.2.2, above) are used to prime the BigDye sequencing reactions. The assembled sequence from the *CS* H7 RealTime amplicon is then used to determine the H7 sample as being either low pathogenic (LP) or highly pathogenic avian influenza (HPAI).

5 Results

5.1 Analysis and Display of Results by Using the RealTime Instrument's Software

5.1.1 Conduct this after familiarising yourself with your RealTime PCR software and instructions.

5.2 Interpretation of Results in Both H7 RealTime PCRs

5.2.1 Analyse the data by comparing the results obtained for the negative (NTC) and H7 positive controls.

5.2.2 Negative Controls (NTCs)

All NTCs should give "No Ct" as their final result. High Cts in all NTC wells with a linear character (e.g. >38) and giving a very low level final fluorescence (ie little greater than the initial "flare" fluorescence values at early cycles) suggest that probe degradation may have occurred, eg the probe has been excessively frozen and thawed. Although such observations may not invalidate the experiment, it is advised to discard the aliquot and thaw-out a fresh aliquot of the relevant H7 probe for subsequent experiments and note result.

If the late Ct value (>38) has a logarithmic/sigmoidal character where clear final fluorescence values are observed, then contamination of the NTC wells with H7 RNA may be considered. It is also possible that such very late Ct signals may occasionally occur spuriously. Repeat the H7 RealTime PCR experiment.

5.2.3 H7 Positive Controls

1. H7 extraction control (3.2.5, Paragraph 2),

Quantified specimens should reproducibly amplify with a Ct value ±2 Cts either side of their predetermined Ct value. Greater increases above the predetermined Ct value suggest that

- RNA extraction from the specimens has been suboptimal. Repeat.

- It is also possible that the H7 positive control may have degraded if stored incorrectly e.g., 4°C, excessive freeze-thawing etc.
- Deviation (±2 Cts) may also occur where software settings on the instrument are inaccurate / inappropriate and the fluorescence cut-off level excessively high / low and is affecting the Ct value of the predetermined H7 positive control.
- It may be possible to reset the software after the experiment to restore the expected Ct value provided the other controls are in order.

2. H7 RNA dilution series "standards" (3.2.5, Paragraph 3)

Ten-fold dilution series should yield a straight line with R2 value of >0.975 (ideally >0.985), slope in the range of –3.0 to –3.9 which should correspond to H5 RealTime PCR efficiency in the range 80–110%.

Deviation from these conditions may not on its own invalidate the test, provided the low positive standards have given Ct's within 2 Cts of the expected value (see Sections 5.2.5 and 5.2.6 below).

However, the results cannot then be interpreted quantitatively. This deviation may reflect eg (i) inaccurate RNA addition (ii) degradation of the RNA standards by inappropriate storage or (iii) inefficient thermo-cycling/fluorescence reading due to an instrument problem. This indicates a need to address these matters for subsequent H7 RealTime PCR experiments.

5.2.4 If all controls are within acceptable limits, analyse the data for the test samples.

5.2.5 Criteria for assessing results on test samples are as follows in the *HA2* H7 Realtime PCR when used on Stratagene Mx3000 series RealTime instruments. The below example Ct values may *not* apply to other makes of RealTime PCR platform:

1. Examine the H7 RNA dilution series. Note the Ct value of the control RNA extracted from a Eurasian H7 sample at a titre of 10 EID_{50}/mL (this Ct being an average of duplicate testing) – refer to Section 3.2.5 Paragraph 3 (above). This should yield a Ct value in the range 34–37 by both H7 RealTime PCRs. In this example, where Ct 36 is obtained from the 10 EID_{50}/mL RNA control, a Ct < 36 in the *HA2* H7 RealTime PCR is considered clearly H7 positive, while a specimen recording "No Ct" value is clearly negative.
2. Samples with Ct >36 in the *HA2* H7 RealTime PCR, but with a logarithmic/sigmoidal character giving clear final fluorescence greater than that observed during early cycles. These are generally considered as "inconclusive", however the following two paragraphs present a possible strategy for confirmation as either "H7 [very low] positive" or "H7 negative"

3. Check the Ct value when the same RNA extract was tested by the M gene RealTime PCR protocol. A Ct value of 34–37 by M gene Real-Time PCR suggests that this may be a very weak positive *HA2* H7 RealTime result. A reproducible result obtained after re-extraction of RNA from the same clinical material by *HA2* H7/M gene RealTime PCRs would tend to confirm this. However, it is unlikely that such "late Ct risers" (particularly those from field submissions) would yield positive results by other tests eg virus isolation or conventional H7 PCRs due to their lower sensitivity.
4. However, if *HA2* H7/M gene RealTime PCR retesting of re-extracted RNA yields a "No Ct" result then such specimens are considered negative.

5.2.6 Criteria for assessing results on test samples are as follows in the *CS* H7

RealTime PCR when used on Stratagene Mx3000 series RealTime instruments.

The example below of Ct values may *not* apply to other makes of Real-Time PCR platform:

1. These criteria are similar in principle to those listed in the above four paragraphs of Section 5.2.5. However, validation with clinical specimens has shown the *CS* H7 RealTime PCR may be less sensitive (up to x10 fold) in comparison to the *HA2* RealTime PCR, and this discrepancy may be even greater for Ct values of >30. Examine the H7 RNA dilution series. The 100 EID_{50}/mL extracted RNA standard typically gives a Ct value of 32 (±2) in the *CS* H7 RealTime. In this example, a Ct <32 in the *CS* H7 RealTime PCR is considered clearly H7 positive, while a specimen recording "No Ct" value is clearly negative.
2. It is possible that *CS* amplification plots (ie with a sigmoidal/logarithmic character) may be observed in the Ct range 32–38. Samples which give such inconclusive results should be re-tested and/or reconsidered with the results of another RealTime PCR (eg M gene and/or HA2 H7) as outlined for the HA2 version, Section 5.2.5, Paragraphs 2–4 (above).

5.2.7 All H7 positive results from a potential new case of AI are repeated and for an index case should be confirmed by use of virus isolation in accord with current OIE/EU guidelines. For the latter conventional H7 PCR and sequencing before reporting is crucial in order to confirm the H7 subtype and pathotype.

5.2.8 Other considerations for the two H7 RealTime PCRs.
Both H7 RealTime PCRs described in this protocol were designed, optimised and validated to detect recent Eurasian H7 isolates. Because of

clear sequence differences between American and Eurasian H7 AIVs, it was not considered likely the former H7s would be detected reliably by the two H7 RealTime PCRs described in this protocol. Results from the validation work affirmed this supposition.

5.2.9 Primer and probe considerations for Eurasian H7 isolates
As Eurasian H7 viruses continue to evolve, it is possible that new isolates will emerge which will be less efficiently detected by the two described H7 RealTime PCR methods, or possibly not detected at all. Revision of H7 primer/probe sequences may be necessary and this will be reflected in any future updated versions of this protocol.

6 Bibliography

1. Spackman, E, Senne, DA, Myers, TJ, Bulaga, LL, Garber, LP, Perdue, ML, Lohman, K, Daum, LT, Suarez, DL. 2002. Development of a real time reverse trancriptase PCR for type A influenza virus and the avian H5 and H7 haemagglutination subtypes. *J. Clin. Microbiol.*, **40**, 3256–60.

SOP 3. One Step RT PCR for Detection of H5 & H7 Avian Influenza and Cleavage Site Sequencing

SOP VI.545 03/09/07

This protocol is a copy of the standard operating procedure used by the avian influenza CRL at the Veterinary Laboratories Agency.

If you have any technical queries please contact aiwrl@vla.defra.gsi.gov.uk

1 Introduction

1.1 Purpose/Scope of This Protocol

This protocol provides details required to determine and analyse the nucleotide sequence of samples of H5 & H7 avian influenza (AI) submitted for statutory sequencing. PCR conditions outlined below include those which have been successful in generating amplicons from RNA directly extracted from clinical specimens.

1.2 Background Information

1.2.1 As described in the OIE diagnostic manual, determination of the nucleotide sequence (and the deduced amino acid sequence) of specific regions of the genomes of these viruses allows an estimation of their pathogenicity to be made.

2 Safety

2.1 Local Safety Procedures

It is your laboratory's responsibility to ensure all work described in this protocol is conducted to a high safety standard. This includes an awareness of risks relating to e.g., dangerous or toxic chemicals, potentially hazardous procedures etc. Here local safety rules in your laboratory should be understood by all relevant members of staff.

3 Procedure/Method

3.1 RNA Extraction

This should be carried out (i) manually or (ii) robotically.

3.2 OneStep RT PCR (Qiagen Kit)

3.2.1 The protocol below is designed round the OneStep RT-PCR kit (Qiagen, cat # 210212).
Prepare the master mix for 50 µL reaction volumes where (i) cDNA is synthesised followed by (ii) PCR in the same tube.

Reagent	Final concentration	Vol (µL) Required for One Reaction (50µL)	Total (µL)
RNase-free water	–	28.8 µL	
PCR Buffer 5X from Qiagen OneStep RT-PCR kit	1X	10 µL	
dNTPs Mix 10mM each (from Qiagen kit)	0.4mM each	2 µL	
Forward primer: 50 pmol/µL (50µM)	1 µM	1 µL	
Reverse primer: 50 pmol/µL (50µM)	1 µM	1 µL	
RNase Inhibitor 40 U/µL (Promega)	8 U	0.2 µL	
One Step RT-PCR Enzyme Mix (Qiagen kit)		2 µL	
Volume minus target		45 µL	
Volume extracted RNA		5 µL for clinical specimens, but 2.5 µL RNA plus 2.5 µL RNase-free water should suffice for egg-grown AI	
Final reaction volume		50 µL	

3.2.2 Optional Variation
Addition of "Q-Solution" has been suggested by the manufacturer for the above kit protocol (Qiagen) to assist in the amplification of difficult templates/specimens. Here 10 µL (×5) Q-Solution can be included into each 50 µL reaction volume. The manufacturer's protocol shows examples where inclusion of Q-Solution can sometimes help overcome problems in generating H5 AI amplicons, but recommends that reactions both with and without Q-Solution should be run in parallel.

3.3 H5 PCR Primers

3.3.1 It is important to note the following when working with these two conventional H5 PCR approaches. Results from VLA and from collaborating EU laboratories in the AVIFLU project have revealed the following:

- H5KHA PCR

Produces approx 300 bp amplicon. This appears to be highly sensitive and successfully amplifies H5 RNA directly extracted from clinical specimens. However, some non-H5 clinical samples can produce a band(s) of similar mobility, where amplicon sequencing is the only means of establishing whether these are derived from H5 or other microbes.

- J3/B2a PCR

Produces approx 300 bp amplicon. This is less sensitive than the KHAH5 PCR, but can produce H5 amplicons from higher-titre clinical specimens. H5 specificity and general robustness appears to be better than for the KHAH5 PCR conditions. Preferable to use this PCR rather than H5 KHA (above) when amplifying RNA extracted from H5 AI which has been harvested from egg fluid.

- S1/B2a PCR

Produces approx 1200 bp amplicon which spans the HA1 region including the cleavage site sequence. Normally used to amplify H5 RNA extracted from egg-grown H5 AI, but may also amplify from very high titre clinical specimens.

3.3.2 H5 Primer Sequences

H5-kha-1 CCT CCA GA**R** TAT GC**M** TA**Y** AAA ATT GTC
H5-kha-3 TAC CAA CCG TCT ACC AT**K** CC**Y** TG

Note, the inclusion of degenerate nucleotides indicated above in bold.

J3 GAT AAA TTC TAG CAT GCC ATT CC

B2a TTT TGT CAA TGA TTG AGT TGA CCT TAT TGG

S1 AGC AGG GGT ATA ATC TCT C

B2a As above

3.4 H7 PCR Primers

3.4.1

- GK7.3/GK7.4: Produces an approx 200–230 bp amplicon which spans the cleavage site. These are the most sensitive primer pair which can amplify from H7 clinical specimens.
- 397/391 and H7.5/391. These two primer pairs produce an approx 600 and 1100 bp amplicons respectively. These span the HA1 region including the cleavage site sequence. Normally used to amplify H7 RNA extracted from egg-grown H7 AI, but may also amplify from very high titre clinical specimens.

3.4.2 H7 Primer Sequences

397	ACATACAGTGGGATAAGAACC
391	TCTCCTTGTGCATTTTGATGCC
GK7.3	ATGTCCGAGATATGTTAAGCA
GK7.4	TTTGTAATCTGCAGCAGTTC

3.5 Cycling Conditions and Amplicon Detection

3.5.1 This Should Be Carried Out as Follows for the Two Primer Pairs

H5KHA cycling conditions

30 min at 50°C
94°C for 15 min
40 cycles of 94°C for 30 s
58°C for 1 min and 68°C for 2min
and a final extension at 68°C for 7 min.

NB: Note the above caveat concerning possible non-H5 amplicons with this H5KHA PCR.

J3/B2a cycling conditions
30 min at 50°C
95°C for 15 min
35 cycles of 94°C for 45 s
50°C for 45 s and 72°C for 2 min
and a final extension at 72°C for 10 min.

S1/B2a; 397/391; GK7.3/GK7.4 & H7.5/391 cycling conditions: precisely as for J3/B2a PCR above.

3.5.2 Electrophorese in 2–2.5% Agarose and Detect by Ethidium Bromide Staining

3.6 Nucleotide Sequencing

This should be carried out by a method of choice in your laboratory, eg ABI BigDyes, Beckman etc.

- Respective primers used for H5/H7 PCR as above, and/or
- In the case of the larger amplicons (ie S1/B2a and H7.5/391), use additional primers within the amplicon which can provide sequence data for both the cleavage site and the HA1 region of the respective H5/H7 genes

For H5: S1/B2a amplicon would require J3 and B2a for sequencing

For H7: H7.5/391 amplicon would require 397 and 391, and /or GK7.3 and GK7.4 for sequencing.

4 Results

4.1 The nucleotide sequence data should be analysed using appropriate sequence analysis software of your choice, eg Lasergene (DNAstar). The common cleavage site nucleotide sequences (and amino acid motifs), and points for their interpretation are listed in Appendix 1.

4.2 Sequencing Results, Fitness for Purpose

Where fluorescent dye sequencing chemistry is employed, the sequencing chromatograms which span the cleavage site region should be assessed critically. Two examples are shown below. The top chromatogram represents a single sequencing run (ie in one orientation) where in places the quality of the data is poor (ie superimposed peaks of more than one signal, a general background pattern of red (ie T) signals), as seen from the character of the plot. This results in clearly one ambiguous nucleotide call at the cleavage site, and likely errors among neighbouring nucleotides which upon translation results in poor aminoacid sequence data at the cleavage site. Such a result would be considered "unfit for purpose" unless there was a complementary sequencing run available (ie in the opposite orientation) of clearly superior quality.

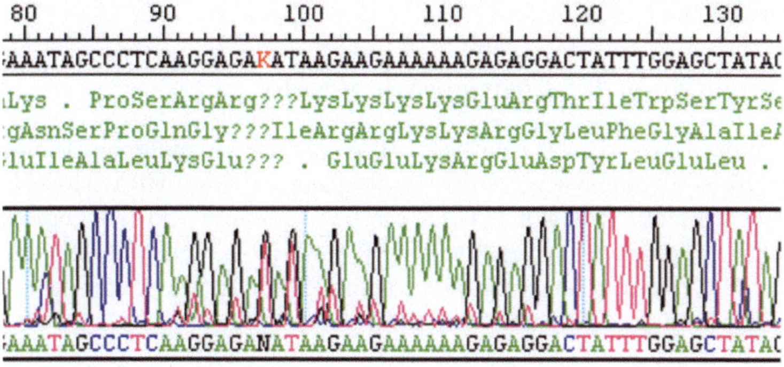

- The next example (below) consists of two complementary sequencing runs in opposite orientations. Compare the quality of the two below chromatograms with the above, ie clear and discrete peaks which agree in both orientations. There are no superimposed peaks nor is there any underlying background, and translation into amino acid sequence is perfectly reliable in giving a polybasic highly pathogenic AI cleavage site in this example. Clearly the below data is "fit for purpose".

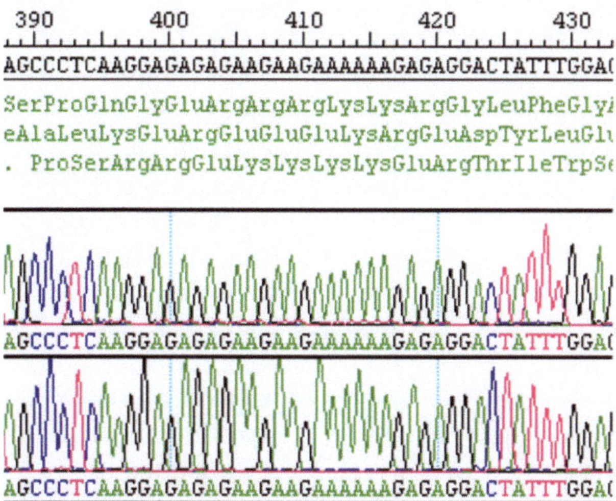

5 Bibliography

1. Slomka, MJ, Coward, VJ, Banks, J, Löndt, BZ, Brown, IH, Voermans, J, Koch, G, Handberg, KJ, Jørgensen, PH, Cherbonnel-Pansart, M, Jestin, V, Cattoli, G, Capua, I, Ejdersund, A, Thorén, P, Czifra, G, 2007a. Identification of sensitive and specific avian influenza PCR methods through blind ring trials in the European Union. *Avian Dis.* **51**, 227–34.

Appendix 1

Interpretation of H5 Cleavage Site Results

An avian influenza virus is considered HPAI if there are multiple basic (argenine – R or lysine- K) amino acids at the HA cleavage site. Most HPAI virus HAs have the sequence motif R-X-R/K-R-G (X can be any amino acid) although there are exceptions. In practice any motif with more than two basic amino acids should be considered HPAI or to be at risk of becoming HPAI. The cleavage site always begins with a proline – P and ends with RGLF.

Table 5.1 Examples of H5 AI amino acid sequences across the cleavage site (not inclusive)

H5 cleavage site motifs			
Virus	**Subtype**	**Motif**	**Pathogenicity**[a]
Numerous LPAI Eurasian viruses	H5N*	PQRET----RGLF	–
A/turkey/Kfar Vitkin/71	H5N2	PQREA----RGLF	–
A/turkey/Ramon/73	H5N2	PQREA----RGLF	–
A/laughing gull/AK/296/75	H5N3	PSIGE----RGLF	–
A/arctic tern/AK/300/75	H5N3	PSIGE----RGLF	–
A/mallard/Alberta/271/88	H5N3	PEKQT----RGLF	–
A/chicken/Italy/RA9097/98	H5N9	PQKET----RGLF	–
A/chicken/Scotland/59	H5N1	PQRKK----RGLF	+
A/tern/South Africa/61	H5N1	PQRETRRQKRGLF	+
A/turkey/Ontario/7732/66	H5N9	PQRRKK---RGLF	+
A/chicken/PA/1370/83	H5N2	PQKKK----RGLF	+
A/duck/Ireland/113/83	H5N8	PQRKRKK--RGLF	+
A/turkey/Ireland/1378/83	H5N8	PQRKRKK--RGLF	+
A/turkey/England/91	H5N1	PQRKRKT--RGLF	+
A/chicken/Pueblo/94	H5N2	PQRKRKT--RGLF	+
A/chicken/Queretaro/20/95	H5N2	PQRKRKRKTRGLF	+
A/chicken/Hong Kong/990/97	H5N1	PQRERRRKKRGLF	+
A/Hong Kong/156/97	H5N1	PQRETRRKKRGLF	+
A/Hong Kong/486/97	H5N1	PQRRRRKK-RGLF	+
A/poultry/Italy/97	H5N2	PQRRRKK-RGLF	+
A/chicken/Italy/1487/97	H5N2	PQRRKKR-RGLF	+
A/chicken/Thailand/04 (1, 2, 3)	H5N1	PQRERRRKKRGLF	+
A/avian/Texas/04	H5N2	PQRKKRGLF	+
A/ostrich/ZA/N227/04	H5N2	PQREKRRKKRGLF	+

[a]Pathogenicity of viruses as determined by IVPI or equivalent tests. + = HPAI, – = LPAI

* Tables 5.1 and 5.2 demonstrates the similarities and differences between be H5 and H7 avian influenza isotypes ranging from low pathogenicity to high pathogenicity. This illustrates that single nucleotide changes within the cleavage site can result in a pathogenicity change for a particular virus isolate. Both H5 and H7 isotypes are reportable to OIE

Table 5.2 Examples of H7 AI amino acid sequences across the cleavage site (not inclusive)

Virus	Subtype	Motif	Pathogenicity[a]
Numerous American viruses	H7N*	PENPK----TRGLF	–
Numerous American and Eurasian	H7N*	PEIPK----GRGLF	–
A/ruddy turnstone/NJ/65/85	H7N3	PEKPK----TRGLF	–
A/duck/OH/10/88	H7N8	PESPK----TRGLF	–
A/finch/CA/28710-8/93	H7N8	PEIPK----ERGLF	–
A/pekin robin/CA/30412-5/94	H7N1	PEIPK---RRRGLF	–
A/FPV/Brescia/02	H7N1	PSKKR---KKRGLF	+
A/FPV/Dobson/27	H7N7	PELPK-KRRKRGLF	+
A/FPV/Dutch/27	H7N7	PPKKR---RKRGLF	+
A/FPV/Rostock/34	H7N1	PEPSK-KREKRGLF	+
A/FPV/Weybridge	H7N7	PELPK-KRRKRGLF	+
A/FPV/Egypt/45	H7N7	FSKKR---RKRGLF	+
A/turkey/England/63	H7N3	PETPK--RRRRGLF	+
A/chicken/Victoria/76	H7N7	PEIPK-KKEKRGLF	+
A/duck/Victoria/76	H7N7	PEIPK----KRGLF	–
A/turkey/England/199/79	H7N7	PEIPK-KREKRGLF	+
A/chicken/Leipzig/79	H7N7	PEIPK--KKGRGLF	+
A/goose/Leipzig/192/79	H7N7	PEIPKKKKKGRGLF	+
A/chicken/Victoria/1/85	H7N7	PEIPK-KREKRGLF	+
A/chicken/Pakistan/447/94	H7N3	PETPK-RKRKRGLF	+
A/chicken/Pakistan/547-99/95	H7N3	PETPK--RRNRGLF	+
A/chicken/Pakistan/1369/95	H7N3	PETPK--RRKRGLF	+
A/Teal/Taiwan/19. 2-37-2/98	H7N1	PEIPK-GRGLF	
A/poultry/Italy/1999	H7N1	PEIPKGSRVRRGLF	+
Italy/7159-V02/02	H7N3	PEIPKGRGLF	–
Pakistan/various/01	H7N3	PETPK--RRKRGLF	+
Pakistan/27/01	H7N3	PEIPKGRGLF	-
Pakistan/33/ 01	H7N3	PETPKRRKRGLF	+
A/poultry/Chile/2002	H7N3	PEKPKTCSPLSRCRETRGLF	+
A/poultry/Chile/2002	H7N3	PEKPKTRGLF	–
A/chicken/NL/2003	H7N7	PEIPKRRRRGLF	+
A/turkey/NL/2003	H7N3	PEIPKGRGLF PEIPKGSRVRRGLF	– +
A/ck/Pakistan/03	H7N3	PETPK-RRKRGLF	+

SOP 4. Eurasian H5 Avian Influenza Real Time PCR

SOP VI.492 Edition 4 18/09/07

This protocol is a copy of the standard operating procedure used by the avian influenza CRL at the Veterinary Laboratories Agency.

If you have any technical queries please contact aiwrl@vla.defra.gsi.gov.uk

1 Introduction

1.1 Purpose/Scope of This Protocol

This protocol provides the details required to detect Eurasian isolates of H5 avian influenza (AI) by RealTime PCR.

1.2 Background Information

This is a rapid, sensitive and specific method for the detection of H5 AI. Validation at VLA has revealed it to be effective in detecting H5 AI directly from clinical specimens, i.e. there is no need to first isolate the suspect H5 isolate in embryonated eggs by growth to a high titre (biological amplification). H5 RealTime PCR has been described previously [1], but validation was restricted to a largely N American collection of H5 AI isolates. For the method described in this protocol, haemaglutinnin gene sequences from Eurasian H5 isolates (mainly 1996–2004) were aligned. Here the H5 primer and hydrolysis probe sequences of Spackman et al. [1] were modified to ensure effective detection of recent Eurasian H5 isolates.

2 Safety

It is your laboratory's responsibility to ensure all work described in this protocol is conducted to a high safety standard. This includes an awareness of risks relating to e.g., dangerous or toxic chemicals, potentially hazardous procedures etc. Here local safety rules in your laboratory should be understood by all relevant members of staff.

3 Materials

3.1 Worksheet for RealTime Plate Layout

Instruction manual for RealTime PCR platform and software in use in your laboratory.

4 Chemicals and Reagents

4.1 Primers and Probe (Note Degenerate Nucleotides Shown in Bold)

H5LH1	*ACA TAT GAC TAC CCA CA**R** TAT TCA G*
H5RH1	*AGA CCA GCT A**Y**C ATG ATT GC*
H5PRO	FAM-TC**W** ACA GTG GCG AGT TCC CTA GCA-TAMRA

4.1.1 H5 RealTime PCR Master Mix

This utilises the Qiagen OneStep RT-PCR kit (Cat No. 210212). Reagents not contained in the kit can be obtained from other suppliers. Volumes indicated below will be sufficient for 10 × 25 µL reactions, i.e. divide the Master Mix into 10 × 23 µL volumes and add 2 µL extracted RNA to each.

DEPC treated water (ie RNAse-free)	Ambion or similar	137.5	µL
(x5) Qiagen 1 step RT PCR buffer		50	µL
Rox ref dye (pre-diluted 1:500 from Stratagene stock in DEPC water) Stratagene or similar		3.75	µL
Qiagen dNTP mix		10	µL
H5LH1 (50 µM)	Sigma or similar	2	µL
H5RH1 (50 µM)	Sigma or similar	2	µL
H5PRO (30 µM)	Sigma or similar	2.5	µL
25 mM magnesium chloride	Promega or similar	12.5	µL
RNAsin (40 U/µL)	Promega or similar	1	µL
Qiagen 1 step RT PCR enzyme mix		10	µL

4.1.2 H5 Controls

Negatives
Include at least two "no template controls" (NTCs) where 2 µL RNase free water is added to each well.

Positives

Quantified H5 specimen e.g., aliquot of inactivated freeze-dried H5N A/chicken/Scotland/59 should serve as an extraction control (manual or robotic). The Ct value of this extracted RNA has been predetermined earlier, and ideally use in duplicate for each H5 RealTime experiment. Enter as "unknown" (or equivalent) during experimental set-up using the software on the RealTime instrument.

Use any other H5 RNA preparation from extracted egg fluid which has been diluted into a x10 fold dilution series (at least 4 dilutions). Ideally this should be a duplicate dilution series. This series (entered appropriately as "standards" with relative quantities indicated by using the software while setting up the RealTime experiment) can be used to determine the H5 RealTime PCR efficiency.

5 Procedure/Method

5.1 Preparation of PCR Master Mix and Loading

5.1.1 Preparation of PCR master mix and loading of real time plate/strip(s) to be carried out in the PCR clean room.

5.1.2 Make up master mix sufficient for the number of samples to be tested.

5.1.3 Thoroughly mix the master mix and centrifuge for 30 s to remove bubbles.

5.1.4 Aliquot 23 μL of master mix per reaction in the RealTime plate/strip(s).

NB: Ensure the plasticware (i.e., plates/strips and covers/caps) is appropriate for the chosen RealTime PCR platform.

5.1.5 Loosely place the plate/strip caps on the plate/strip(s). All wells will need to be covered even though only a portion of the plate may be used.

5.1.6 Bring the plate/strip(s) out of the clean room and place on ice before addition of sample RNA and controls.

5.2 Preparation of H5 RNA Controls

5.2.1 H5 positive controls are as described above in Section 3.2.3.

5.2.2 Extract RNA as per the manufacturer's protocol

5.2.3 Include H5 positive controls (ideally in duplicate) in each H5 RealTime experiment.

5.3 Manual Addition of Samples and Standards

5.3.1 Regardless of whether RNA extraction has been conducted manually or robotically (VI 537), at VLA we manually add 2 μL of the sample RNA to the 23 μL of master mix. A dedicated multichannel pipette is convenient for adding robotically extracted RNA.
Take care to avoid cross contamination of samples at this stage. Change gloves frequently and do not hold tips containing RNA above the incorrect wells.

5.3.2 Add 2 μL of RealTime PCR controls to appropriate wells based on your worksheet layout. Re-apply caps to the rest of the plate/strip(s).

5.3.3 Once RNA is added, fit caps to all wells.
It is important that the caps are fitted firmly and correctly onto the wells before insertion into the RealTime instrument.

5.3.4 If the RealTime plate/strip(s) are not to be loaded into the instrument immediately, keep the plate/strip(s) on ice until ready to test.

5.4 Reverse Transcription (RT) and PCR

5.4.1 Place the real time plate/strip(s) in the appropriate real time machine.

5.4.2 Incubate the reactions with the following thermocycling profile

RT step	50°C for 30 min
	95°C for 15 min
PCR step (x 40 cycles)	95°C for 10 s
	54°C for 30 s
	72°C for 10 s

NB. Collect fluorescence data at end of 54°C step using the ROX and FAM filters.

5.4.3 Open a new file using the instrument software to define all the necessary RealTime PCR experimental parameters. This is conducted according to the manufacturer's software instructions. Provide the file an appropriate H5 RealTime PCR name.

5.4.5 The complete run takes approximately 2 h, but this may vary according to the make of instrument. If the machine is not booked for immediate use by another colleague upon completion of the run, select the option (if available) to turn off the lamp at the end of the run. This is because certain makes of lamp have a limited lifespan.

6 Results

6.1 Analysis and Display of Results

6.1.1 Conduct this after familiarising yourself with your RealTime PCR software and instructions.

6.2 Interpretation of Results

6.2.1 Analyse the data by comparing the results obtained for the negative (NTCs) and H5 positive controls.

6.2.2 Negative Controls (NTCs)

All NTCs should give "No Ct" as their final result. High Cts in all NTC wells with a linear character (e.g. >38) and giving a very low level final fluorescence (i.e. little greater than the initial "flare" fluorescence values at early cycles) suggest that probe degradation may have occurred, e.g., the probe has been excessively frozen and thawed. Although such observations may not invalidate the experiment, it is advised to discard the aliquot and thaw-out a fresh aliquot of H5 probe for subsequent experiments and note result.

If the late Ct value has a logarithmic/sigmoidal character where clear final fluorescence values are observed, then contamination of the NTC wells with H5 RNA must be considered. Repeat the experiment.

6.2.3 H5 Positive Controls

Quantified specimens that serve as an extraction control (e.g. A/chicken/Scotland/59 (H5N1), Section 3.2.3) should reproducibly amplify with a Ct value ±2 Cts either side of their predetermined Ct value of e.g. 30. Greater deviations from the predetermined Ct value suggest that:

- The H5 positive control may have degraded (if Ct greater than predetermined value).
- Software settings on the instrument are inaccurate/inappropriate and the fluorescence cut-off level excessively high/low and is affecting the Ct value of the predetermined H5 positive control. It may be possible to reset the software after the experiment to restore the expected Ct value provided the other controls are in order.
- Deviation (±2Cts) may also occur where software settings on the instrument are inaccurate/inappropriate and the fluorescence cut-off level excessively high/low and is affecting the Ct value of the predetermined H7

positive control. It may be possible to reset the software after the experiment to restore the expected Ct value provided the other controls are in order.

Ten-fold dilution series of RNA should yield a straight line with R^2 value of 0.975 (ideally >0.985), slope in the range of –3.0 to–3.9 which should correspond to H5 RealTime PCR efficiency in the range 80–110%.

6.2.4 If all controls are within acceptable limits analyse the data for the test samples.

6.2.5 Criteria for assessing results on test samples are as follows

1. *Samples with Ct <33* H5 positive.
2. *Samples with Ct >33* but with a clear *logarithmic/sigmoidal character* giving clear final fluorescence greater than that observed during early cycles. At the time of writing these are considered inconclusive by H5 RealTime PCR. However, it is important to consider the results of other tests e.g. virus isolation in embryonated eggs (followed by typing with reference antisera) and/or H5 conventional PCRs and sequencing which may reveal a H5 positive result for the specimen.
3. *Samples with Ct >33* but with a *linear character* giving final fluorescence which is approximately the same as that observed during early cycles.
   a) As noted above in Section 5.2.2, such a result may be due to H5 probe degradation. Resort to using fresh aliquot of probe.
   b) Alternatively, other possible explanations include inefficient hybridisation at late cycles to non-specifically amplified nucleic acid of host origin in the RNA sample.
   c) Such results are most likely to be H5 negative, but consider the results of other tests e.g. virus isolation in embryonated eggs (followed by typing) and/or H5 conventional PCRs to exclude a H5 positive result for the specimen.
4. *No Ct value*, H5 RealTime PCR negative.

6.2.6 All H5 positive results from a potential new case of AI are repeated and for an index case should be confirmed by use of virus isolation in accord with current OIE/EU guidelines. For the latter conventional H5 PCR and sequencing before reporting is crucial in order to confirm the H5 subtype of detected AI.

6.2.7 Other considerations for the Eurasian H5 RealTime PCR

This H5 RealTime PCR was designed and validated to detect recent Eurasian H5 isolates. However, there is some evidence that certain American lineage H5s may be detected by this procedure. However, it must be stressed that detection of American H5 isolates by this Eurasian H5 RealTime PCR has not been validated.

6.2.8 Primer and probe design for Eurasian H5 isolates

As these viruses continue to evolve, it is possible that new isolates will emerge which will be less efficiently detected by the above method, or even not detected at all. Revision of primer/probe sequences will be necessary and this will be reflected in any future updated versions of this protocol.

7 Bibliography

1. Slomka, MJ, Pavlidis, T, Banks, J, Shell, W, McNally, A, Essen, S, Brown, IH. 2007. Validated H5 Eurasian real-time reverse transcriptase–polymerase chain reaction and its application in H5N1 outbreaks in 2005–2006. *Avian Dis.*, **51**, 373–7.

SOP 5. Detection of Rift Valley Fever Virus by Real-Time Reverse Transcription-PCR

Courtesy of Janusz T. Paweska,
Head Special Pathogens Unit,
National Institute for Communicable Diseases of the National Health Laboratory Service,
Private Bag X4, Sandringham-Johannesburg, 2131,
Modderfontein Road 1, Gauteng,
SOUTH AFRICA

Responsible scientist: Janusz Paweska
E-mail: januszp@nicd.ac.za

1 Detection of Rift Valley Fever Virus by Real-Time Reverse Transcription -PCR

1.1 Equipment

Micropipettes	10 μL, 100 μL, 1000 μL
Micropipettes filter tips	10 μL, 100 μL, 1000 μL
Microcentrifuge	
Biohazard flow cabinet	
LightCycler instrument	
Ice machine	

2 RNA Isolation

2.1 Material and Reagents

QIAamp viral RNA Mini kit
LightCycler RNA Amplification Kit HybProbe
Primers
Probe

2.2 In Extraction Room

For the extraction of RNA from
Plasma
Serum
Cell- free body fluids or Cell culture supernatants- use the QIAamp viral RNA Mini kit according to the instructions of the manufacturer (QIAgen, Germany, catalogue number 52906, 250 reactions).

2.3 Primers and Taqman Probe

Primers 5′–3′

RVS (sense primer) **5′ AAAggAACAATggACTCTggTCA 3′**

RVAs (antisense primer) **5′ CACTTCTTACTACCATgTCCTCCAAT 3′**

Probe 5′-nuclease probe labeled with 6-carboxyfluorescein at the 5′end and with 6-carboxy-N,N,N′,N′-tetramethylrhodamine at the 3′ end.

RVP 5′ – 6FAM – AAAgCTTTgATATCTCTCAgTgCCCCAA --TMR

3 RT-PCR Method

3.1 In PCR Mix Room

Use the LightCycler RNA Amplification Kit HybProbe (Roche, catalogue number 12 015 145 001, 96 reactions).

In a 1.5 mL reaction tube on ice, prepare the PCR mix for one 20 μL reaction by adding the following components

Component	Volume	Final conc.
LightCycler RT-PCR reaction mix HybProbe 5x conc. (vial 2)	5 μL	1x, MgCl 3mM
Mg_2Cl stock solution (vial 3)	2 μL	2 mM
Primer RVS (10 mM)	1.25 μL	0.5 μM
Primer RVAs (10 mM)	1.25 μL	0.5 μM
Probe RVP (10 mM)	0.5 μL	0.2 μM
RT-PCR enzyme	0.5 μL	
Water (vial 4)	9.5	
Total volume	20 μL	

To prepare the PCR Mix for more than one reaction, multiply the amount in the "Volume" column above by z, where z = the number of reactions to be run + one additional reaction

- Mix gently by pipetting up and down. Do not vortex.
- In the PCR mix room, pipet 20 μL PCR mix into each LightCycler Capillary.
- In the extraction room, add 5 μL RNA template. Seal each capillary with a stopper.
- Place the adapters (containing the capillaries) into a standard benchtop microcentrifuge.
- Place the centrifuge adapters in a balanced arrangement within the centrifuge.
- Centrifuge at 700×*g* for 5 s (3000 rpm in a standard benchtop microcentrifuge).
- Transfer the capillaries into the sample carousel of the LightCycler Instrument.
- Cycle the samples as described below.

Program for lightcycler		
Reverse transcription		
Cycles	1	
Analysis mode	None	
Temperature targets	**1**	
Target temperature	45°C	
Incubation time (h:min:s)	00:30:00	
Temperature transition rate	20°C/s	
Secondary target temperature	0	
Step size	0	
Step delay (cycles)	0	
Acquisition mode	None	
Denaturation		
Cycles	1	
Analysis mode	None	
Temperature targets	**1**	
Target temperature	95°C	
Incubation time (h:min:s)	00:05:00	
Temperature transition rate	20°C/s	
Secondary target temperature	0	
Step size	0	
Step delay (cycles)	0	
Acquisition mode	None	
Amplification	**Signal analysis F1/F2**	
Cycles	45	
Analysis mode	Quantification	
Temperature targets	**1**	**2**
Target temperature	95°C	57°C
Incubation time (h:min:s)	00:00:05	00:00:35
Temperature transition rate	20°C/s	20°C/s
Secondary target temperature	0	0
Step size	0	0
Step delay (cycles)	0	0
Acquisition mode	None	Single
Cooling		
Cycles	1	
Analysis mode	None	
Temperature targets	**1**	
Target temperature	30°C	
Incubation time (h:min:s)	00:00:30	
Temperature transition rate	20°C/s	
Secondary target temperature	0	
Step size	0	
Step delay (cycles)	0	
Acquisition mode	None	

SOP 6. Swine Vesicular Disease (SVD) Virus One-Step RT-LAMP

Courtesy of Mikhayil Hakhverdyan, Anne-Lie Blomström
National Veterinary Institute,
Department of Virology,
Ulls väg 2B
SE-751 89 Uppsala
SWEDEN

Responsible scientist Mikhayil Hakhverdyan
E-mail: mikhayil.hakhvedyan@sva.se

Created 2008-02-24
Updated 2008-02-25

1 Introduction

Loop-mediated isothermal amplification (LAMP) is a method used to amplify RNA and DNA. Two or three primer pairs are used to amplify the template which gives a long stem- loop product under isothermal conditions. This report describes a one-step reverse transcriptase (RT) LAMP assay for the sensitive (50 viral RNA copies per assay) and rapid detection (within 30–60 min) of swine vesicular disease virus (SVDV). The results can be visualized either by gel electrophoresis or by the naked eye through the addition of SYBR®Green. For more information about the assay development and validation see Blomström et al., 2008, *J. Virol. Methods*, **147,** 188–193. This RT-LAMP assay provides a number of benefits for the diagnosis of SVD, since the assay is sensitive and rapid, and the isothermal amplification strategy used is not reliant upon expensive equipment it is particularly suited for “front line” diagnosis of SVD in modestly equipped laboratories, in field stations or in mobile diagnostic units.

2 RNA Extraction

Manual TRIZOL (Sigma, USA) extraction of SVDV isolates, which were of cell culture origin, was used according to the manufacturer’s instructions. The clinical samples (200 µL of each) were extracted using MagNA Pure LC Total Nucleic Acid Isolation Kit (Roche, Mannheim, Germany) according to the manufacturer’s protocol and eluted in 50 µL.

3 RT-LAMP

3.1 General Aspects

a) The highly conserved 3D polymerase gene was chosen as target for the primer design.
b) The RT-LAMP reaction is carried out in a 25 µL mixture.
c) After production of the master mix, 23 µL portions are filled into each reaction tube and 2 µL template are added.

3.2 Master Mix Preparation

a) The volume of the master mix depends on the number of extracted samples plus the number of positive and negative controls (water).
b) Before dividing up the master mix into portions of 23 µL per tube, the master mix must be homogenised thoroughly.
c) The RT-LAMP reaction is carried out in a 25 µL mixture containing 1× Thermo buffer (New England Biolabs, Beverly, MA, USA), 0.54× First strand buffer (Invitrogen, Carlsbad, CA, USA), 1.12 mM dNTPs (GE Healthcare, Uppsala, Sweden), 0.2 µM each of F3 and B3, 1.6 µM each of FIP and BIP, 0.8 µM each of Floop and Bloop, 0.8 µM betaine (Sigma), 4.5 U cloned AMV reverse transcriptase (Invitrogen, Carlsbad, CA, USA) and 8U Bst DNA polymerase, Large Fragment (New England Biolabs).

3.3 Addition of Template RNA

a) 2 µL template RNA is added to each tube with 23 µL master mix, except for the negative control for which water is added.
b) After addition of template RNA, close the reaction tubes and run LAMP.

3.4 Thermodynamic Profile

Reverse Transcription/incubation	63°C	60 min
Inactivation	80°C	2 min

3.5 Analysis of Results

a) Run 5 μL of the RT-LAMP product on a 1.5% agarose gel.
b) Digest 2 μL product with BspHI (New England Biolabs) for 1 h at 37°C before analysis by gel-electrophoresis to confirm the result.
c) Add 0.5 μL 10× SYBR®-Green Gel Stain I 10,000 × concentration (Molecular Probes, Leiden, the Netherlands) to 10 μL RT-LAMP product to visualised the result. After a short vortex a colour change can be seen if the sample is positive.

4 Appendix Primer Mixes

F3	*ACC AAT TCA GAA TGA TTG CA*
B3	*GGA TGG ACC AAA AAY GGG TA*
FIP	*TCC CTG CTT CAG CTA GCA GTT TTT GAT GAT GTG ATA GCT TCA TAC C*
BIP	*CAG CAG ATA AAG GCG AGT GTT TTT TTT CAT CTG CCC TAA ART ACC T*
Floop	*AGG CAT CGA TRG GCC ACG*
Bloop	*ACC TGG ACA AAC GTG ACC T*

5 Abbreviations

F3 – forward primer
B3 – backward primer
FIP – forward internal primer
BIP – backward internal primer
Floop – forward loop primer
Bloop – backward loop primer

Protocol

Transmittal /laboratory number ………………… Transmittal date……………………
RT-LAMP run number …………………..… …. Date……………………………..……
Executing person …………………………….. Time…………………………
Running file saved …………………….………………………………….
Comments ……………………………..…………………………………..

Pipetting steps	Master mix components	Colour code	SVDV RT-LAMP	
			1 x	N x
1.	RNase free water		2.2 μL	
2.	10× Thermo buffer		2.5 μL	
3.	5× First strand buffer		2.7 μL	
4.	10mM dNTPs		2.8 μL	
5.	F3 primer (10 μM)		0.5 μL	
6.	B3 primer (10 μM)		0.5 μL	
7.	FIP primer (20 μM)		2 μL	
8.	FIP primer (20 μM)		2 μL	
9.	Bloop primer (10 μM)		2 μL	
10.	Floop primer (10 μM)		2 μL	
11.	Betaine (5 μM)		2.5 μL	
12.	Cloned AMV reverse transcriptase (15 U)		0.3 μL	
13.	Bst DNA polymerase, Large Fragment (8000 U/ml)		1 μL	
	Master mix total volume		**23 μL**	
6.	**Addition of template RNA** Field samples NTC (add RNase free water)		per 2 μL	
	Total reaction volume		**25 μL**	

Plate assignment

Temperature profile Results…………………......
RT/incubation 60 min 63°C ………………………………….
Inactivation 2 min

SOP 7. Detection of African Swine Fever Virus DNA Using the Isothermal

ASFV vp72 655T Invader Squared Assay

LAB-ON-SITE
EU PROJECT
SSPE-CT-2004-513645

Courtesy of Bernt Hjertner

The Queen's University of Belfast,
Department of Veterinary Science,
Stoney Road, BT4 3SD Belfast,
United Kingdom.

1 Scope of SOP

This SOP describes the sensitive and specific detection of African swine fever virus DNA using the isothermal ASFV vp72 655T Invader squared assay. The Invader™ assay is a linear, isothermal (63°C) signal amplification system able to accurately quantify DNA and RNA targets with high sensitivity and specificity.

The assay is a good alternative to PCR as it can be used with simple and inexpensive equipment. This particular procedure describes the ASFV Invader assay and its use in conjunction with the iQ™5 multicolor real-time PCR detection system (Bio-Rad). It has a detection limit around 2500 copies. The somewhat lower sensitivity as compared to PCR is compensated by the possibility to use it as an end point assay. Possible targets could include extracted genomic material, cDNA, existing PCR product or crude/purified plasmid.

The kit described in this protocol, ASFV VP72 655T, include vials contains

VP72 655T Oligo Mix (0.25 µM Invader, 2.5µM Probe)

Cleavase XI Enzyme, FRET mix

Control target VP72 655T

No Target Control (tRNA).

For more information on the chemistry behind the tecchnology and the development of the assay please consult the Instruction sheet "INVADER™ ASSAY: VP72 655T 345 detection from viral DNA Fluorescence Resonance Energy Transfer

(FRET) Detection Format (Third Wave Agbio) and Hjertner et al. 2005 *J. Virol. Methods* **124**, 1–10, respectively.

2 Samples and DNA Extraction

The samples were extracted using the to "Boom-silica nucleic acid extraction protocol" developed by Partner 3, DTU Vet, Lindholm, but other methods can be used.

Include following controls

Extraction controls

Negative control sample material, NCS, processed the same way as the clinical samples.

Positive spiked control sample material, SCS, (negative control sample spiked with positive control oligo).

3 Isothermal Reaction

3.1 General Aspects

a) The Cleavase enzyme is not a hot start enzyme so the reactions have to be kept on ice the whole time. The master mix should not be left sitting for long as this can increase background.
b) Include the following Invader controls from the kit. No target blank and positive control oligo (NTB and PCO).
c) After making the master mix 7.5 μL portions are filled into each reaction tube and 7.5 μL template is added.

3.2 Master Mix Preparation

a) The volume of the master mix depends on the number of extracted samples plus the number of controls and possible standards.
b) Before dividing up the master mix into portions of 7.5 μL per well, the master mix must be homogenised thoroughly.

3.3 Addition of Template DNA

a) Template DNA is boiled for 10 min and chilled on ice. To each well containing 7.5 μL master mix another 7.5 μL template DNA is added.

b) For quantification of DNA a standard curve must be generated. Six dilutions of positive genomic ASFV DNA are produced in the range 1×10^6 to 1×10^3 copies/ µL.
c) After addition of template DNA mix thoroughly and overlay with 15 µL spectroscopically clear mineral oil.
d) Close the reaction tubes and spin down the PCR preparation.

3.4 Cycler Program

The programme consists of a 4 h incubation at 63°C measuring fluorescence every 2.5 min.

Read Fluorescence channel FAM.
If not using a real-time instrument incubates in heatblock/water bath and measure fluorescence every hour.

3.5 Analysis of Generated Fluorescence Data

a) Ensure that NCS is negative and SCS is positive with a slope value or end point read (see separate SOP) similar to PCO.
b) If NCS or SCS deviate from this repeat the extraction.
c) Ensure that NTB is negative but PCO is positive with an expected slope value or end point read.
d) If the NTB or PCO control deviates repeat the Invader reaction.
e) A sample is considered positive if the slope value or end point data obtained can be statistically differentiated from the NTB using Students TTEST (see separate SOP).
f) The analysis of field samples is possible if all controls respond as expected.
g) Select all standards and analyse the slope rates or end point data as described in separate SOP. -
h) If a linear standard curve is achieved with an R-squared value of 0.95 or greater use it to quantify the clinical samples.

4 Appendix. Oligos

Invader oligo 5′ TGGTTGTCCCAGTCATATCCGTTGCGAGGAT 3′
Probe oligo 5′ acggacgcggagAACGTTTGAAGCTGACC 3′
Positive control oligo
5′CCATGGTCAGCTTCAAACGTTTCCTCGCAACGGATATGACTGGGACAACCAAA 3′

The underlined 5′ end of the probe oligo (=FLAP, is released by target specific cleavage) is complementary to a generic FRET cassette provided with the kit. This cassette when cleaved generates the fluorescence.

Protocol

Transmittal /laboratory number .... Transmittal date....................

Real-time PCR run number.................................Assay.......................

Executing person.............. Date...........

Running file saved...................... T ime...........

Comments..

Pipetting steps	Master mix components	Volume	
		1x	**Nx**
1.	Oligo mix (0.25 μM Invader, 2.5 μM Probe)	3.0 μL	
2.	FRET mix (FRET cassette)	3.5 μL	
3.	Cleavase XI Enzyme	1.0 μL	
	Master mix total volume	**7.5 μL**	
4.	**addition of template DNA** Field samples, NCS, SCS Standards, NTB, PCO	7.5 μL	
	Total reaction volume	**15 μL**	

Plate assignment

Temperature profile

1 cycle	equilibrating	1 min	63°C
99 Cycles	soak	2.5 min	63°C

Results...

SOP 8. Real-Time PCR Detection and Quantification of Porcine Viruses Using Molecular Beacons

LAB-ON-SITE
EU PROJECT
SSPE-CT-2004-513645

Courtesy of John McKillen

The Queen's University of Belfast
Department of Veterinary Science
Stoney Road, BT4 3SD Belfast
United Kingdom

1 Introduction

This SOP describes the sensitive and specific detection of Aujeszky's disease virus (ADV), African swine fever virus (ASFV), porcine circovirus type 2 (PCV2) and porcine parvovirus (PPV).

The assays target the gD gene (ADV), the 9GL gene (ASFV), the Rep gene (PCV2) and the VP2 gene (PPV), amplifying PCR products of 110, 264, 263 and 130 bp respectively.

Each assay has a detection limit of 20 copies. They have been evaluated on whole blood, serum and tissue samples. For more information about the assay development and evaluation with clinical samples see McKillen et al. 2007. *J. Virol. Methods* **140**, 155–65.

2 Samples and DNA Extraction

For details of tested samples see McKillen et al. 2007. *J. Virol. Methods* **140**, 155–65. Purify nucleic acid from whole blood, serum or 10% tissue homogenates in MEME using the Qiagen DNA blood mini Kit according to the manufacturer's instructions.

 # Include following controls

Extraction controls

Negative control sample material, NCS, processed the same way as the clinical saples.

Positive spiked control sample material, SCS, (negative control sample spiked with relevant positive amplicon).

3 Real-Time PCR

3.1 General Aspects

a) The amplification is based on the JumpStart Taq mix (SIGMA-Aldrich) with a total reaction volume of 25 μL.
b) Include the following PCR controls, Negative and positive control (NC and PC).
c) After making of the master mix 23 μL portions are filled into each reaction tube and 2 μL template is added.

3.2 Master Mix Preparation

a) The volume of the master mix depends on the number of extracted samples plus the number of controls and possible standards.
b) Before dividing up the master mix into portions of 23 μL per tube, the master mix must be homogenised thoroughly.

3.3 Addition of Template DNA

a) Two μL template DNA are added to each tube with 23 μL master mix.
b) For quantification of DNA in field samples, a standard curve must be generated. Six dilutions of positive control DNA are produced in the range 1×10^6 to 1×10^1 copies/μL.
c) After addition of template DNA, close the reaction tubes and spin down the PCR preparation.

3.4 Cycler Program

The cycling conditions are optimised to be used on the MJ research Opticon®2 real time thermal cycler.

Activation of Taq	95°C	2 min	
Denaturation	94°C	30 s	40 cycles
Annealing	x°C	30 s	40 cycles
Elongation	72°C	30 s	40 cycles

Read Fluorescence channel FAM (ADV, ASFV, PCV2) or HEX (PPV).
Fluorescence data are collected in the annealing phase.

3.5 Analysis of Generated Fluorescence Data

a) Ensure that NCS is negative and SCS is positive with a CT value similar to PC.
b) If NCS or SCS deviate from this repeat the extraction.
c) Ensure that NC is negative but PC is positive with an expected CT value.
d) If the NC or PC control deviates repeat the PCR.
e) A sample is considered positive if a CT value is obtained and the sample shows a typical sigmoidal curve.
f) If the CT value is lower than the detection limit of the assay used additional tests are recommended.
g) The analysis of field samples is possible if all controls respond as expected.
h) Select all standards and analyse the standard curve for linearity and efficiency according to guidelines of the manufacturer of the PCR platform.
i) If a linear standard curve is achieved with a PCR efficiency of 95% or greater use it to quantify the clinical samples.

4 Appendix. Primers, Probes and Reaction Conditions

Primers

ADV gd F429-445	CGCACCACGCCGATGTG
ADV gd R519-538	CGGTACTGGCCCTCGTTGAA
ASFV 9GL F77	CGGGAGACGTTGTTTTAT
ASFV 9GL R340	CGCCTTTTCGTATCTTAC
PCV GrFor 283-300	CTTCTGCGGTAACGCCTC
PCVII R529Match GFor	TTACCCTCCTCGCCAACA
PPV VP2 F112-132	GGGTTGGTGTGTCTACAGGT
PPV VP2 R190-210	GTATGAGTCTTGATGCGTGTGC

Probes

ADVgd MB462-25	cgctcgGACTACATGTTCCCCACGGAGGACGcgagcg
ASFV 9GL MB162-25	cgctcgGGCTCCAATAAAGTCGGTTTTCCCAcgagcg
PCV2 MB25 394-418	gcgagcCACCTCAGCAGCAACATGCCCAGCAgctcgc
PPV VP2 MB156-25	cgctcgTCAATACTTGGGGGAGGGCTTGGTTcgagcg

Reaction conditions

	T_a(°C)	$Mg2^+$ (mM)	F primer conc (μM)	R primer conc (μM)	Beacon conc (μM)
ADV	50	4	0.75	0.75	0.5
ASFV	51	5.5	0.5	0.5	1.0
PCV2	50	4	0.5	0.5	0.5
PPV	52	4	0.5	0.5	0.5

Primers, size and T_a for standard amplicons

Amplicon	Primer sequence	T_a(°C)	size (bp)
ADV F375–394a	TTTATCGAGTACGCCGACTG	52	248
ADV R604–622a	CGGGCGAACGGGCACTCTT		
ASFV 9GL F77	CGGGAGACGTTGTTTTAT		264
ASFV 9GL R340	CGCCTTTTCGTATCTTAC		
PCV2 SG 181F	AAGATGCCATTTTTCCTT	50	689
PCV GROUP 852-871R	CCCGCTCACTTTCAAAAGTT		
PPV VP2 F112-132	GGGTTGGTGTGTCTACAGGT	52	130
PPV VP2 R190-210	GTATGAGTCTTGATGCGTGTGC		

Protocol

Transmittal /laboratory number.................... Transmittal date.......................

Real-time PCR run number............................ Assay......................................

Executing person.. Date......................................

Running file saved.................................. Time..

Comments...

Pipetting-steps	Master mix components	volume 1 x	volume N x
1.	PCR grade water	To 23 µL	
2.	2x SIGMA jumpstart master mix	12.5 µL	
3.	10 µM Molecular beacon probe	1.25 µL (ASFV: 2.5 µL)	
4.	100 µM primermix	0.125 µL (ADV: 0.1875 µL)	
5.	25 mM $MgCl_2$	.... µL	
	Master mix total volume	**23 µL**	
6.	**addition of template DNA** Field samples, NCS, SCS Standards, NC, PC	2 µL	
	Total reaction volume	**25 µL**	

Plate assignment

Temperature profile

Activation/denaturation	2min	95°	40 Cycles
Denat.	30 s	94°C	
Annealing	30 s	...°C	
Elongation	30 s	72°C	

Results...

....

SOP 9. Swine Vesicular Disease Virus PriProET Two-Step Real-Time PCR

LAB-ON-SITE
EU PROJECT
SSPE-CT-2004-513645

Courtesy of Mikhayil Hakhverdyan
National Veterinary Institute
Department of Virology
Ulls väg 2B
SE-751 89 Uppsala
SWEDEN

Responsible scientist Mikhayil Hakhverdyan
E-mail: mikhayil.hakhverdyan@sva.se

Created 2008-02-15
Updated 2008-02-18

1 Introduction

This SOP describes a two-step protocol for PriProET real-time PCR amplification of swine vesicular disease viruses (SVDV).

The assay amplifies the 3D-gene of any of SVDV strain while heterologous virus strains including Coxsackievirus B5 remain negative.

The sensitivity of assay is five copies of viral genome equivalents. A key point of the assay is tolerance toward mutations in the probe region. Melting curve analysis directly after PCR, with determination of probe melting point, confirms specific hybridisation of the SVDV strains.

More details can be found in the published article, Hakhverdyan, M, et al. 2006. *Arch. Virol.* **151** (12), 2365–76.

The protocol was optimised using ABI 7700 instrument (Applied Biosystems, USA), but later was adopted for RotorGene (Corbett Research, Australia). Both instruments works well with the system, but RotorGene, like many other modern real-time PCR machines, has friendly user format and simplifies analysis of results.

2 RNA Extraction

Manual TRIZOL extraction according to the manufacturer's instructions was used in the current protocol, but any isolation method that gives good RNA yield could be used instead.

3 cDNA Synthesis

cDNA was produced in 25 μL reaction volume using 5 μL RNA, 1 μL random hexamers (pdN6, 0.02 U, Amersham, Uppsala, Sweden), 2.5 μL dNTPs (2 mM), 5 μL 5X First Strand Buffer, 1 μL RNAguard (1000 U, Amersham Bioscience, USA), and 1 μL M-MLV reverse transcriptase (200 U, Ambion, Austin, Texas, USA). The reaction was incubated for 5 min at 22°C, followed by 90 min at 37°C, and the enzyme was finally inactivated by heating for 5 min at 95°C.

4 Real-Time PCR

4.1 General Aspects

a) The amplification is based on the TITANIUM Taq DNA polymerase kit (Clontech) with a total reaction volume of 25 μL.
b) Controls, PC (positive control), NTC (no template control)
c) After production of the master mix, 23 μL portions are aliquoted into each reaction tube and 2 μL of template is added.

4.2 Master Mix Preparation

a) The volume of the master mix depends on the number of extracted samples plus the number of controls (PC; NTC) plus one-two extra reactions.
b) Before dividing up the master mix into portions of 23 μL per tube, the master mix must be mixed thoroughly and pulse spanned for 5 s.

4.3 Addition of Template cDNA

a) 2 μL template cDNA is added to each tube with 23 μL master mix and close the reaction tubes.
b) Spin down reaction mix (optional).

4.4 Cycler Program

Preliminary denaturation	–95°C	2 min	
Denaturation	95°C	15 s	55 cycles
Annealing	60°C	15 s	
Elongation	72°C	15 s	
Denaturation	95°C	15 s	
Probe melting profile		96 cycles of [50°C for 10 s with auto-increments of 0.5°C] for ABI 7700 or 40°C →95°C for RotorGene.	

Fluorescence channel is FAM for ABI 7700. Filter set for RotorGene is 470nm/610hp (emission/detection)
Fluorescence data are collected in the annealing phase and during the probe melting profile

4.5 Analysis of Generated Fluorescence Data

General aspects

a) A Ct value of approx. 25 should be detectable for PC and a specific probe melting point of 57.5–72.5°C (depends on the number of mutations in the probe region). Each mutation in the probe region decreases melting point on 5°C.
b) No Ct value or melting point should be detectable for NTC.
c) The analysis of field samples is possible if all controls respond as expected.
d) A suspected SVDV sample is considered as positive in the real-time PCR, if a Ct value is detected for the sample and/or if the FAM fluorescence increases significantly over the base level.
e) For confirmation of a positive SVDV specific real-time PCR result the probe melting profile has to display a specific probe melting point of 70°C±2.5°C corresponding to a perfect probe match, while T_m =65±2.5°C and T_m =60±2.5°C corresponding to one or two mutations, respectively.

5 Appendix. Primer /Probe Mixes

SVDV-3D-6511-F-FAM (10 µM)	5′ FAM-TCAACCCGGGCATCGTTAC 3′
SVDV-3D-6617-R (10 µM)	5′ TGAATAGTCAAACGCTATGAGATGTC 3′
SVDV-3D-6553-P-probe (10 µM)	5′ GGGTCACACCCAACGGCGCT-Texas Red 3′

Predicted amplicon size is 107 bp

Protocol in ABI7700 or RotorGene 3000/6000

Transmittal /laboratory number ……………………..…Transmittal date………….............

Real-time PCR run number…………………Date……………………….............

Executing person…………………………………..…Time……………………..............

Running file saved………………… ………………………………….……………

Comments……………………………..…………………………………………......……

Pipetting steps	Master mix components	Colour code	SVDV PriProET	
			1 x	N x
1.	DNase free water		16.25 µL	
2.	10X TITANIUM Taq PCR buffer		2.5 µL	
3.	4X dNTPs		1.25 µL	
4.	SVDV-3D-6511-F-FAM (10 µM)		1.25 µL	
5.	SVDV-3D-6617-R (10 µM)		0.5 µL	
6.	SVDV-3D-6553-P-probe (10 µM)		0.75 µL	
7.	50X TITANIUM Taq DNA Polymerase		0.5 µL	
	Master mix total volume		**23 µL**	
8.	**Addition of template cDNA** - Field samples, - PC, NTC (add DNase free water)		per 2 µL	
	Total reaction volume		**25 µL**	

96-well plate or 36/72-well rotor assignment

Temperature profile (amplification)

Activation/denaturation 2min 95°C

55 Cycle
- **Denaturation** 15 s 95°C
- **Annealing** 15 s 60°C
- **Elongation** 15 s 72°C

Temperature profile (melting curve)
- **Denaturation** 15 s 95°C
- **Probe Melting Profile** 96 cycles of [50°C for 10 s with auto-increments of 0.5°C] for ABI 7700 or 40°C->95°C for RotorGene.

The melting curve is obtained directly following the amplification

SOP 10. Slope/End Point Analysis of Invader Data System

LAB-ON-SITE EU PROJECT
SSPE-CT-2004-513645

Courtesy of Bernt Hjertner

The Queen's University of Belfast,
Department of Veterinary Science,
Stoney Road, BT4 3SD Belfast,
United Kingdom.

1 Introduction

This SOP describes the analysis of invader data generated using either a real-time PCR instrument or an ordinary fluorescent plate reader.

As there exist no platform with built in capability to analyze any data generated using an Invader based protocol it has to be extracted and analyzed manually.This can be based on either *reaction slopes* (from real-time data) or *end point* measurements.

This protocol describes the calculations/manipulations necessary to get a quantified and statistically verified analysis of any sample. This is done using the Excel template "DNA invader analysis template" available by request from b.hjertner@qub.ac.uk.

For further information on Invader assays please refer to the Instruction sheet "INVADER™ ASSAY VP72 655T 345 detection from viral DNA Fluorescence Resonance Energy Transfer (FRET) Detection Format (Third Wave Agbio) and Hjertner et al. 2005. *J. Virol. Methods* **124**, 1–10.

2 Data Extraction

a) For data retrieval please refer to the manual describing the platforms used to generate the data.
b) Paste the data in an empty excel sheet so that cycle numbers start in column A and continue down, the samples starting in column B and onward.

3 Determining Slope Rate

Mark column B, right-click and chose insert. In the inserted column add the time points.

a) Copy the entire row (A) in the top outlining your wells and paste the whole row under the last data collected. Under this row write the label of each sample.
b) Usually, only the data range corresponding to up to 10% of the strongest scoring sample is used for each sample. This is chosen as follows. Place the cursor between the brackets in the "10%MAX" box. Delete the × by hitting backspace. Now mark the entire data field and hit enter.
c) Go back and mark the entire data field again and chose conditional formatting under format. Choose "Cell value is" and "greater than or equal to" and in the third box double click followed by clicking on the 10%MAX value. Then click on format and under color chose red. Click on ok and again ok. All data points higher than 10% of the strongest data point are now marked red.
d) Now start a slope rate row under your sample definition row. For the first well write "= slope(,)". Place the cursor before the comma and mark all black data points for that sample. Likewise, place the corresponding time points after the comma. Repeat this for all the samples.

3.1 Statistical Analysis (Slope Rates)

a) Go to the sheet named slope analysis. This template contains an analysis matrix for duplicate samples and ten known standards.
b) In the matrix the following are calculated. Average, Standard deviation (SD), Coefficient of variation (CV), Fold over 0, Net signal, Ttest to neighbour and Ttest to 0.
c) If more than duplicates are run mark the row outlining Average, right click and hit insert. Repeat this for every extra repetition.
d) If more than 10 samples are analyzed copy the area comprising the values and all calculations and place the curser in the first adjacent empty square. Right click and choose "paste special". Use the option "formulas" when pasting.
e) Check all squares containing calculations to make sure that the new repetitions are included and that they all link to the right square. Adjust if necessary.
f) Now go to raw data and copy the slope rate for one sample. Go back to slope rate analysis tab and mark its position in the matrix. Right click and in "paste special" choose "values". Click ok. Repeat this for all samples.

3.2 Making a Standard Curve (Slope Analysis)

a) Mark the Net signal values and click on the table icon in the toolbar selecting "XY (scatter)" followed by Next.
b) Choose the tab series and put cursor in the X values box and then mark the copy number series in the matrix.
c) Click Next and in the following display put in chart title and values for X and Y axis. Click Next followed by finish.
d) Place cursor over one of the data points and right click followed by "add trendline". Choose linear in "type" and display R value and equation in "options".
e) Use the equation to quantify any unknown samples.

4 Statistical Analysis (End Point Analysis)

a) Find the row displaying the time point of interest. Right click the entire row and copy it. Paste the row below, under the row of samples.
b) Next steps are identical to slope end analysis. Follow description from 3.1.a. It is not necessary to use the paste special.

4.1 Making a Standard Curve (End Point Analysis)

a) This procedure is identical to slope end analysis (3.2.a onwards).

SOP 11. African Horse Sickness TaqMan RT-PCR

LAB-ON-SITE EU PROJECT
SSPE-CT-2004-513645

Courtesy of Belen Rodriguez

Universidad Complutense de Madrid
Departamento de Sanidad Animal
Facultad de Veterinaria
Avda Puerta de Hierro s/n
28040 Madrid
Spain

Responsible scientist: Belen Rodriguez
E-mail: lbelenrodriguez@sanidadanimal.inf

Created 2008-02-18
Updated 2008-02-18

1 Introduction

This SOP describes the detection by one-step RT-PCR with a TaqMan probe of all the serotypes of AHSV. The primers target a conserved region within the NS1 segment. The PCR product expands 210 bp. The set of primers and the FAM-labelled probed used in this protocol has been designed by Partner 4, UCM, Madrid. We use Trizol Reagent for RNA extraction.

2 RNA Extraction

RNA isolation is performed with Trizol Reagent, following the manufacturer's instructions.

3 Real-Time RT-PCR

3.1 General Aspects

a) The amplification is based on the Brilliant® QRT-PCR Master Mix One-Step kit (Stratagene) with a total reaction volume of 25 µL.
b) Controls, PC (positive control), NTC (no template control).
c) After production of the master mix, 23 µL portions are aliquoted into each reaction tube and
d) 2 µL of template is added.

3.2 Master Mix Preparation

a) The volume of the master mix depends on the number of extracted samples plus the number of controls (PC; NTC) plus two extra reactions.
b) Before dividing up the master mix into portions of 23 µL per tube, the master mix must be mixed thoroughly.

3.3 Addition of Template RNA

a) 2 µL template RNA is added to each tube with 23 µL master mix.
b) After addition of template RNA, close the reaction tubes and spin down the solution.

3.4 Cycler Program

Reverse Transcription	48°C	30 min	
Inactivation RT /Activation Taq	95°C	10 min	
Denaturation	95°C	30 s	} 40 cycles
Annealing-Elongation	58°C	30 s	

Fluorescence channels are ROX (negative control) and FAM.

Fluorescence data are collected in the annealing phase and during the probe melting profile

3.5 Analysis of Generated Fluorescence Data

General aspects
ROX channel can be used for normalisation of the FAM fluorescence
FAM channel

A Ct value of approx. 20 should be detectable for PC
No Ct value should be detectable for NTC

The analysis of field samples is possible if all controls respond as expected.
A suspected AHSV sample is considered positive in the real-time RT-PCR, if a Ct value is detected for the sample and/or if the FAM fluorescence increases significantly over the base level.

4 Appendix. Primer /Probe Mixes

AHSV-F (20 μM)	5'GTTGACCTCGCTCTGCTTGAC3'
AHSV-R (20 μM)	5'GTTGACCTCGCTCTGCTTGAC3'
TAQ_AHSV (probe) (20 μM)	5'FAM-GTTGACCTCGCTCTGCTTGAC-BHQ3'

Protocol in Stratagene Mx3000p

Transmittal /laboratory number ……………… Transmittal date……………............

Real-time RT-PCR run number……………….Date……………………...............

Executing person……………………………… Time…………………….............

Running file saved……………… …………………………….…………………...

Comments…………………………….………………………………………....……

Pipetting steps	Master mix components	Colour code	Pan-SIV PriProET	
			1 x	**N x**
1.	Brilliant QRT-PCR Master Mix		12.5 µL	
2.	PCR-quality water		9.2125 µL	
4.	AHSV-F (20 µM)		0.625 µL	
5.	AHSV-F (20 µM)		0.3 µL	
6.	TAQ_AHSV (20 µM)		0.3 µL	
7.	Enzyme Mixture		0.0625 µL	
	Master mix total volume		**23 µL**	
8.	**Addition of template RNA** Field samples, PC, NTC (add RNase free water)		per 2 µL	
	Total reaction volume		**25 µL**	

Plate assignment

Temperature profile (amplification)

RT	30min	48°C
Activation/denaturation	10 min	95°C
40 Cycle Denaturation	30 s	95°C
Annealing-Elongation	30 s	58°C

SOP 12. Bluetongue SYBR®-Green RT-PCR

LAB-ON-SITE EU PROJECT
SSPE-CT-2004-513645

Courtesy of Belen Rodriguez
Universidad Complutense de Madrid
Departamento de Sanidad Animal
Facultad de Veterinaria
Avda Puerta de Hierro s/n
28040 Madrid
Spain

Responsible scientist : Belen Rodriguez
E-mail: belenrodriguez@sanidadanimal.info

Created 2008-02-18
Updated 2008-02-18

1 Introduction

The SOP described here allows the detection of most of BTV serotypes reported in the Mediterranean basin by one-step RT-PCR.

So far, a 77bp within the NS1 segment belonging to serotypes 1, 2, 4, 8 and 9 have been successfully amplified. The amplification of BTV16 with the primers described here is still ongoing. The set of primers used in this protocol has been designed by Partner 4, UCM, Madrid.

2 RNA Extraction

RNA isolation is performed with Trizol Reagent, following the manufacturer's instructions.

3 Real-Time RT-PCR

3.1 General Aspects

a) The amplification is based on the SYBR® Green QRT-PCR Master Mix One-Step kit (Stratagene) with a total reaction volume of 25 µL.
b) Controls, PC (positive control), NTC (no template control)
c) After production of the master mix, 23 µL portions are aliquoted into each reaction tube and 2 µL of template is added.

3.2 Master Mix Preparation

The volume of the master mix depends on the number of extracted samples plus the number of controls (PC; NTC) plus two extra reactions.

a) Before dividing up the master mix into portions of 23 µL per tube, the master mix must be mixed thoroughly.

3.3 Addition of Template RNA

a) 2 µL template RNA is added to each tube with 23 µL master mix.
b) After addition of template RNA, close the reaction tubes and spin down the solution.

3.4 Cycler Program

Reverse Transcription	48°C	30 min	
Inactivation RT /Activation Taq	95°C	2 min	
Denaturation	95°C	30 s	40 cycles
Annealing	56°C	30 s	40 cycles
Elongation	72°C	30 s	40 cycles
Denaturation	95°C	5 min	40 cycles
Probe melting profile	60°C->95°C		

Fluorescence channels are ROX (negative control) and SYBR® GREEN.
Fluorescence data are collected in the annealing phase and during the probe melting profile

3.5 Analysis of Generated Fluorescence Data

General aspects

ROX channel can be used for normalisation of the SYBR® Green fluorescence

SYBR® Green channel

a) A Ct value of approx. 20 should be detectable for PC and a specific probe melting point of 78°C should be observed.
b) No Ct value or melting point should be detectable for NTC.
c) The analysis of field samples is possible if all controls respond as expected.
d) A suspected BTV sample is considered positive in the real-time RT-PCR, if a Ct value is detected for the sample and/or if the SYBR® Green fluorescence increases significantly over the base level.
e) For confirmation of a positive BTV specific real-time PCR result, the probe melting profile has to display a specific probe melting point of 75–78°C.

4 Appendix. Primer /Probe Mixes

BTVNS1-F (20 μM)	5′ CACATTCACCGCTGGATTAGAG 3′
BTVNS1-R (20 μM)	5′ CGTCCAAGCATGAAAATACC 3′

Protocol in Stratagene Mx3000p

Transmittal /laboratory number ………………….. Transmittal date………….............
Real-time RT-PCR run number………….. Date…………………….............
Executing person………………………………… Time……………………...............
Running file saved………………. ………………………………….………………..
Comments…………………………………………………………………………….................

Pipetting steps	Master mix components	Colour code	Pan-SIV PriProET 1 x	Pan-SIV PriProET N x
1.	SYBR® Green QRT-PCR Master Mix		12.5 µL	
2.	PCR-quality water		9.4375 µL	
4.	BTVNS1-F (20 µM)		0.5 µL	
5.	BTVNS1-R (20 µM)		0.5 µL	
6.	Enzyme Mixture		0.0625 µL	
	Master mix total volume		**23 µL**	
7.	**addition of template RNA** Field samples, PC, NTC (add RNase free water)		per 2 µL	
	Total reaction volume		**25 µL**	

Plate assignment

Temperature profile (amplification)

RT	30 min	48°C
Activation/denaturation	2 min	95°C
40 Cycle		
Denaturation	30 s	95°C
Annealing	30 s	56°C
Elongation	30 s	72°C

Temperature profile (melting curve)

Denaturation	5 min	95°C
Probe Melting Profile	60°C->95°C	

The melting curve is obtained directly following the amplification

SOP 13. BTV Serotype 4 SYBR® GREEN RT-PCR

LAB-ON-SITE
EU PROJECT
SSPE-CT-2004-513645

Courtesy of Belen Rodriguez
Universidad Complutense de Madrid
Departamento de Sanidad Animal
Facultad de Veterinaria
Avda Puerta de Hierro s/n
28040 Madrid
Spain

Responsible scientist Belen Rodriguez
E-mail belenrodriguez@sanidadanimal.info

Created 2008-02-18
Updated 2008-02-18

1 Introduction

This SOP describes a one-step protocol for SYBR® Green RT-PCR amplification of bluetongue virus (BTV).This new RT-PCR has been developed by Partner 4, UCM, Madrid, for the specific detection of BTV serotype 4. The pair of primers described here target a sequence within the VP2 segment which contains deletions only described in BTV4 sequence. The PCR product is 124 bp long.

2 RNA Extraction

RNA isolation is performed with Trizol Reagent, following the manufacturer's instructions.

3 Real-Time RT-PCR

3.1 General Aspects

a) The amplification is based on the SYBR® Green QRT-PCR Master Mix One-Step kit (Stratagene) with a total reaction volume of 25 μL.
b) Controls, PC (positive control), NTC (no template control)
c) After production of the master mix, 23 μL portions are aliquoted into each reaction tube and 2 μL of template is added.

3.2 Master Mix Preparation

a) The volume of the master mix depends on the number of extracted samples plus the number of controls (PC; NTC) plus two extra reactions.
b) Before dividing up the master mix into portions of 23 μL per tube, the master mix must be mixed thoroughly.

3.3 Addition of Template RNA

a) 2 μL template RNA is added to each tube with 23 μL master mix.
b) After addition of template RNA, close the reaction tubes and spin down the solution.

3.4 Cycler Program

Reverse Transcription	48°C	30 min	
Inactivation RT /Activation Taq	95°C	5 min	55 cycles
Denaturation	95°C	45 s	55 cycles
Annealing	58°C	45 s	55 cycles
Elongation	72°C	30 s	55 cycles
Denaturation	95°C	5 s	55 cycles
Probe melting profile	60°C->95°C		

Fluorescence channels are ROX (negative control) and SYBR® Green.
Fluorescence data are collected in the annealing phase and during the probe melting profile

3.5 Analysis of Generated Fluorescence Data

General Aspects

ROX channel can be used for normalisation of the SYBR® Green fluorescence

SYBR® Green channel

a) A Ct value of approx. 20 should be detectable for PC and a specific probe melting point of 78°C should be observed.
b) No Ct value or melting point should be detectable for NTC.
c) The analysis of field samples is possible if all controls respond as expected.
d) A suspected BTV sample is considered positive in the real-time RT-PCR, if a Ct value is detected for the sample and/or if the SYBR® Green fluorescence increases significantly over the base level.
e) For confirmation of a positive BTV specific real-time PCR result, the probe melting profile has to display a specific probe melting point of 78°C.

4 Appendix Primer /Probe Mixes

BTVVP2_F (20 μM)	5′TAGATTCCATCCCGGTGATTCT3′
BTVVP2_R (20 μM)	5′AGGTACATTCACTCCACCCTGC3′

Protocol in Stratagene Mx3005p

Transmittal /laboratory number .... Transmittal date.........................

Real-time RT-PCR run number .... Date...

Executing person............................... Time.....................................

Running file saved.................... ....

Comments...

Pipetting steps	Master mix components	Colour code	Pan-SIV PriProET	
			1 x	N x
1.	SYBR® Green QRT-PCR Master Mix		12.5 μL	
2.	PCR-quality water		9.4375 μL	
3.	BTVVP2-F (20 μM)		0.5 μL	
4.	BTVVP2-R (20 μM)		0.5 μL	
6.	Enzyme Mixture		0.0625 μL	
	Master mix total volume		**23 μL**	
7.	**Addition of template RNA** Field samples, PC, NTC (add RNase free water)		per 2 μL	
	Total reaction volume		**25 μL**	

Plate assignment

Temperature profile (amplification) RT 30min @ 48°C
Activation/denaturation; 5min @ 95°
40 Cycle, Denaturation 45 s @ 95°C
Annealing 45 s @ 58°C
Elongation 30 s @ 72°C

Temperature profile (melting curve) Denaturation 5 min 95°C . Probe Melting Profile 60°C @ 95°C
The melting curve is obtained directly following the amplification

SOP 14. Real-Time Duplex Detection of Avian Influenza and Newcastle Disease Viruses

LAB-ON-SITE – EU PROJECT
SSPE-CT-2004-51364

e-mail: istvan.kiss@sva.se or sandor.belak@bvf.slu.se

National Veterinary Institute,
Department of Virology,
Ulls väg 2B
SE-751 89 Uppsala
SWEDEN

1 Introduction

The AIV/NDV duplex LUX (Light upon extension) assay is qualified for the sensitive and specific real-time detection/discrimination of AIV/NDV strains originating from various sources. For the published description of the separate assays see Kiss, I. et al. 2006. *Acta Vet Hung.*, **54**, 525–533., and Antal, M., et al. 2007. *J. Vet. Diagn. Invest.* **19**, 400–04.

The LUX assay does not require oligonucleotide probe and quencher molecule, instead, it uses only two primers – like conventional PCR – one of which is labelled with a fluorophore molecule. This set-up enables melting curve analysis on completion of the amplification, which provides a convenient and reliable way for confirming its specificity. The assay is capable of rapidly detecting a broad range of influenza A and Newcastle disease viruses. Its specificity, sensitivity and cost-efficiency make the assay a useful tool for AI/ND diagnostics.

2 RNA Extraction

RNA is extracted from 140 μL allantoic fluid, 10% (w/v prepared in distilled water) feces suspensions/organ homogenates or lyophilized virus strains reconstituted in distilled water with the QiAamp Viral RNA Mini Kit (Qiagen, Valencia, CA) as recommended by the manufacturer. RNA is eluted in 60 μL of elution buffer and 1 μL per reaction is used as template for the one-step RT-PCR.

3 Real-Time RT-PCR

3.1 Primers Used in the Assay

Forward primer AIV		Reverse primer AIV	
name	sequence (5′–3′)	name	sequence (5′–3′)
AI-For-37	gacctGATGAGTCTTCTAACCGAGG**T**C T (in bold) next to the 3′ end C is labelled with 6-carboxyfluorescein (FAM), the lower case letters indicate the nucleotides added to the target-specific sequences in order to form hairpin	AI-Rev-181	TGTCTTTAGCCAYTCCATGAG
Forward primer NDV		Reverse primer NDV	
name	sequence (5′–3′)	name	sequence (5′–3′)
ND-703F-JOE	catctt AGT GGC AGT TGG GAA GA**T**G the **T** next to the 3′ end G is labeled with the fluorophore JOE, 6-carboxy-4,5-dichloro-2,7-dimethoxyfluorescein; lower case letters indicate the nucleotides added to the targetspecific sequences to form hairpin	ND-845R	GTG GYC CGA ATA CTG TAG
Both PCRs yield an appr. 145 bp PCR product			

3.2 Real-Time RT-PCR

The Qiagen one-step RT-PCR kit (Qiagen, Valencia, CA) is used for the PCR assays in a final volume of 25 μL.

The assay set-up is the following

Component		Cycling temp. (°C)	Time	Function
Name	Amount (μL)	50	60 min	RT
5x buffer	5	95	15 min	Activation
$MgCl_2$ (25 mM)	1	94	15 s	45 cycles for amplification
H_2O	12.6	58	35 s	
Rnase inhibitor (25 U/μL)	0.2	72	30 s	
dNTP (10 mM)	1	72	1 min	
Forward primer (10 μM)	0.6 + 0.6			
Reverse primer (10 μM)	1 + 1			
Enzyme	1			
sample	1			

Melting points are at 88.5 ± 0.5°C 85.5±0.5°C for AIV and NDV, respectively.

In no-template controls distilled water served as target. As internal amplification controls (IAC) pSP73 vector derived (Promega) constructs are used.

In the AIV specific assay, the IAC, which contains plasmid born sequences between the AIV LUX primer binding sites yields a 256 bp PCR product upon amplification.

The same type of IAC is used for NDV but it carries a resistance marker flanked by the NDV LUX primer binding sites and yields a 978 bp PCR product. The analytical sensitivity is appr. 20 plasmid copies for both assays.

Reading of fluorescence for FAM and JOE is performed after the synthesis step. Having completed the PCR a melting point analysis is done on both channels.

The assay was tested on the following instruments, iCycler® Bio-Rad), ABI 7500 real-time PCR system (Applied Biosystems, Foster City, CA), and RotorGene 3000 (Corbett Research, Mortlake, NSW, Australia.

SOP 15. Realtime RT PCR Detection of Influenza Virus Matrix Gene Realtime RT PCR Detection of Velogenic Newcastle Disease Fusion Protein

Courtesy of Sharon Hietala and Beate Crossley
California Animal Health and Food Safety Laboratory
University of California,
Davis
United States of America.
Email: skhietala@ucdavis.edu

The SOP is formatted to include procedural information for performance of realtime RT PCR for surveillance and diagnostic specimens in a single tube or 96-well high-throughput format.

The information includes

1. Establishment and Monitoring of Quality Control Specimen Performance Limits
2. Handling and Storage of realtime PCR Primer and Probes
3. Performance of real time RT PCR, Avian Influenza virus matrix gene and virulent marker/fusion protein for Newcastle disease virus.
   a) Single tube extraction using commercial spin column technology, suitable for testing small numbers of specimens.
   b) Manual high-throughput extraction using commercial 96-well magnetic bead technology. The approach allows a single technician to extract RNA for subsequent PCR analysis from 384 specimens in approximately 60 min without robotic technologies.
4. Guidance document for response to assay failures.

1 Procedure

1.1 Establishment and Monitoring of QC Specimen Performance Limits

1.2 Purpose

The performance ranges obtained for the realtime PCR Quality Control specimens are used to determine the pass/fail status of the diagnostic assays performed, as well as to continuously monitor and improve assay performance. Quantitative values obtained from replicate testing of the Quality Control specimens are recorded with each use, statistically evaluated over time, and performance ranges are updated on a regular basis.

1.3 Scope

This protocol describes the steps used to establish the acceptable performance range of Quality Control specimens used for realtime PCR assays.

1.4 Definitions and Acronyms

Ct value, real time PCR Cycle threshold value.

Armored RNA (AR)

An RNA virus mimic comprised of assay-specific target ribonucleic acid sequences encapsulated in a protein coat.

Quality Control (QC)

Extraction Control

Specimen used to assess extraction efficiency for realtime PCR assays.

The extraction control can be the infectious agent, a clinical specimen containing the agent of interest, or assay-specific AR.

PCR Control

Previously extracted agent-specific nucleic acid stored at –70°C and tested with each assay run to assess performance the PCR components of the realtime PCR assay.

Mean

Sum of data point values divided by number of data points.

Standard Deviation

Quantifies the dispersion of the data point values around the mean, and is used to set limits for acceptable performance ranges. 1 standard deviation from the mean will include 68.3% of the values assuming a normal Gaussian distribution, 2 standard deviations from the mean will include 95.5%, and 3 standard deviations will include 99.7% of the values assuming a normal **Gaussian Distribution.**

Coefficient of Variation (%CV)

An alternate measure of variability expressed as a percentage. The %CV is used for precision (reproducibility) testing of multiple replicates (minimum of 5 replicates) of the same specimen.

1.5 Materials Required

Continuous (numerical) data is recorded using an assay specific tracking system, such as an Excel spreadsheet or equivalent QC tracking form used for recording the mean, standard deviation, and %CV of assay control specimens.

1.6 Procedure/Detail

To establish the acceptable performance range for quality control specimens (Cycle threshold values)

1. Use a minimum of 10 data points from independent assay runs to calculate the mean Ct value for each QC specimen.
   a. For establishing initial QC specimen performance evaluations; calculate %CV and proceed only if the control specimen used has a %CV less than 15. It is permissible to exclude statistical outliers prior to proceeding to next steps, however additional data to allow a minimum of 10 data points for analysis are required.
   b. For assays with previously determined acceptable performance ranges (defined as mean ±2 standard deviations) and target performance range (defined as mean ±1.5 standard deviations), continuously update the performance range every 10–20 runs by adding new QC data to existing data. For the statistical analysis exclude data points from failed runs, defined as assay runs that were due to explained or unexplained assay failure (example, any or a combination of QC specimen realtime PCR Ct values beyond the accepted performance range due to pipetting errors, dilution errors, and etc.). Verify extreme Ct values (outliers) using the original values to identify and correct potential data entry errors (example, QC values range between 25.5 and 28.5Cts, and one of the values was documented as 2.85 Ct).
2. Calculate the mean and standard deviation of the valid data points.
3. Establish the quality control range as ± 2 standard deviations of the mean value. Example
   a. mean value 21.5 Ct
   b. standard deviation 0.89 Ct
   c. Ct quality control assay acceptance range 21.5 ± 0.89 = 19.72 to 23.28 Ct
4. Record the established QC values on Quality Control tracking form for the specific assay.

Note. To assist tracking and updating of QC specimen values, mark the final data entry where the QC range was established (e.g. in the spreadsheet insert a colored row at the last entry used in the calculation), and note the prior acceptable range at this point (mean value ± 2 standard deviations). With continued assay use and additional data points, the acceptable performance range should progressively show less variability (smaller standard deviation) before becoming stabile. The magnitude of assay variability is heavily dependent on original assay design, e.g. lack of primer-dimers, GC content of the target region, other assay interference resulting from choice of nucleic acid sequences, as well as sample matrices used in the assay with tissues generally showing more innate variability than fluids.

Acceptable performance ranges are established, monitored, and updated for QC specimens to include at least,

a) Extraction controls.
b) Armored RNA controls.
c) PCR controls.
d) Negative extraction and PCR control values without armored RNA.
e) Negative extraction and PCR control values with armored RNA.

5. Both negative and positive QC is recorded with every assay performed.
6. QC values from failed are tracked and investigated to assess source and frequency of assay failure, identify failures as random or systematic error, and respond to trends in assay performance (refer to assay failure response guideline).

Exceptions

1. For assays where PCR and Extraction controls are used as matched controls (the PCR control and the Extraction control are prepared from the same initial stock infectious agent dilution and are used together each time the assay is performed to enhance assay between-run reproducibility); additional acceptance criteria is monitored and used to accept assay performance; e.g., the Ct value for the PCR control is valid, the Ct value for the paired Extraction control is valid, *and* the difference between the PCR and Extraction controls remains constant (example, Extraction control CT – PCR control Ct ≤ 3 Ct.).
2. For assays where AR is used as an internal control, a valid range is established for the expected AR Ct values. If AR is added to the negative extraction control preparation, QC performance ranges must be established.
   Note. Armored RNA is added as an internal control to clinical specimens and is optimized to show Ct values near the detection limit of the assay. Where a clinical sample tests strong positive for the target RNA, the armored RNA may test negative due to target-specific PCR competition in the assay. AR negative values obtained in positive clinical specimens should be neglected during the calculation of the acceptable AR Ct range. Note. This situation can be avoided by using weak positive specimen as the extraction control, dependent on the

availability of that specific source material. Document all false AR negative values with Ct of positive sample to update efficiency measure calculations.

3. For assays that are validated for either of two or more extraction methods and are to be compared for performance reporting or assessment of detection limit, mathematical correction must be applied before the QC range is established, e.g. extraction using a spin column (e.g. silica gel column A) uses a different clinical specimen start volume and results in a different elution volume compared to a magnetic bead extraction method (e.g. bead-based method B). The mathematical correction factor must be determined and applied to the Ct values for each new method before the acceptable performance ranges for the QC specimen can be determined.

1.7 Appendices

Mean: $\overline{x} = \sum x / n$, where x is the data value and n is the number of data points.

Standard Deviation (SD)

$$\sigma = \sqrt{\frac{\Sigma\,[X-\overline{X}]^2}{n}}$$

σ = standard deviation

Σ = sum of

X = each value in the data set

$\overline{X}$ = mean of all values in the data set

n = number of value in the data set

Coefficient of Variation (CV%) = (Standard Deviation divided by the mean) times 100.

1.8 Bibliography

1. Donia, D, et al. 2005. Use of armored RNA as a standard to construct a calibration curve for real-time RT-PCR. *J. Virol. Methods*, **126** (1–2), 157–63.
2. Pasloke, BL, et al. 1998. Armored RNA technology for production of ribonuclease-resistant viral RNA controls and standards. *J. Clin. Microbiol.*, **36** (12), 3590–4.
3. Westgard, JO, et al. 1990. Selection of medically useful Quality Control procedures for individual tests done in a multitest analytical system. *Clin. Chem.*, **36**, 230–3.

2 Procedure – Preparation of Stock Primer and Probes

2.1 Scope

PCR assay primer and probe (dye-labeled hydrolysis probes) inventory is maintained in100uM aliquots; 10uM aliquots are prepared from the Stock PCR reagent concentration for assay use and short-term storage (<60days).

2.2 Materials Required

1. Primer and probe are typically obtained in lyophilized form from commercial suppliers.
2. Tris-EDTA (TE) Buffer (10 mM Tris, 1mM EDTA; pH 8.0)
3. Amber-colored microtubes or foil wrapped microtubes to protect probe fluorescent dyes from light-associated decay.

2.3 Procedure/Detail

For 100 μM Stock Solution

1. Use the "*total nmoles*" value supplied by the manufacturer and multiply by *10* to determine the volume of diluent in microliters required, example, "*total nmoles*" × *10* = μL volume of diluent added.

Example, 13.57 nmoles × 10 = 135.7 μL of TE buffer added to tube

Add 135.7 μL TE buffer to tube. Vortex well. Pulse centrifuge to insure no liquid is adherent to the cap.

10 μM Working Solution

1. To make a single 50 μL aliquot of 10 **μM** working solution (1:10 dilution of stock solution), add 5 μL of 100 μM stock solution to 45 μL of TE buffer. Vortex and pulse spin.
2. To make more than one aliquot, dilute in one tube and then aliquot to the remaining tubes. *Example*, To make 6, 50 μL tubes of a 10 μM working solution (300 μL total volume).

   $C_1V_1=C_2V_2$; $V_1 = (C_2V_2)/C_1$

   $C_1 = 100$ μM , $V_1 =$ μL of stock solution
   $C_2 = 10$ μM, $V_2 = 300$ μL
   $V_1 = (C_2V_2)/C_1 = (10 \times 300)/100 = 30$ μL of stock solution

3. Pipette 270 µL of TE buffer in a tube, then add 30 µL of the stock solution. Vortex and pulse spin.
   Aliquot 50 µL into individual storage tubes.
4. Label tube with primer or probe name and tube sequence number. Label each tube with the concentration and date prepared for Quality Assurance and inventory tracking purposes.
5. New lots and working solution of primers and probes must be tested three times side-by-side with the previous lot for Quality Assurance purposes. The Ct results are documented; and performance limits must be verified as within the performance range previously established for each probe and/or primer sets (Mean Ct ± standard deviations).

2.4 Document and Record Requirements

A reagent inventory identifying stock reagents by name, concentration, date prepared, lot number and tube sequence number is maintained for quality assurance and laboratory management purposes. Aliquots associated with documented assay failure that is tracked to lot number, stock concentration aliquots, or working concentration aliquots are destroyed.

3 Procedure- Realtime RT PCR Detection of Influenza Virus Matrix Gene, Realtime RT PCR Detection of Velogenic Newcastle Disease Fusion Protein Gene

3.1 Purpose

The SOP describes the procedures used for realtime PCR detection of the Avian Influenza virus matrix gene and velogenic Newcastle Disease virus fusion protein gene, including RNA recovery followed by realtime PCR amplification and detection performed in a single tube format used for testing small numbers of samples, or in a manual (non-robotic) high throughput 96-well format for testing large numbers of samples.

3.2 Scope

Realtime RT PCR detection of the Influenza virus matrix gene and velogenic Newcastle Disease virus fusion protein gene markers are used for surveillance and diagnostic screening of clinical materials, including oral-pharyngeal swabs, tracheal swabs, tissues, culture media, and allantoic fluids. The Influenza matrix assay is not species-specific, and may be used with avian, swine, equine specimens. The Newcastle disease virus fusion protein target is specific to poultry species.

3.3 Materials Required

Isopropanol, 100%

Ethanol, 100% analytic grade

10% bleach solution, freshly-made, for disinfection/disposal of pipette tips

Influenza Matrix gene forward primer, 5′-AGA TGA GTC TTC TAA CCG AGG TCG-3'
Influenza Matrix gene reverse primer, 5′-TGC AAA AAC ATC TTC AAG TCT CTG-3′
Influenza Matrix gene probe, 5′-FAM-TCA GGC CCC CTC AAA GCC GA-BHQ1-3′
(TaqMan or Molecular Beacon with FAM/Blackhole Quencher labels; Integrated DNA Technology, Inc or equivalent).

Newcastle disease fusion protein gene forward primer,
5′-GGT GAG TCT ATC CGG ARG ARG TA CAA G-3′

Newcastle disease fusion protein gene reverse primer,
5′-AGC TGT TCG AAC CCC AAG-3′

Newcastle disease fusion protein velogenic marker probe,
5′-FAM-AAG CGT TTC TGT CTC CTT CCT CCA-BHQ1-3′

(TaqMan or Molecular Beacon with FAM/Blackhole Quencher labels; *Integrated DNA Technology, Inc* or equivalent).

3.3.1 Single Tube PCR

Spin Column for RNA recovery; *Qiagen Inc*, *Ambion/Asuragen* Inc., *Roche* Inc., or equivalent commercial kit. Kit components are prepared according to the manufacturer's instructions.

Beta-mercaptoethanol
70% Ethanol, molecular grade
100% Ethanol, molecular grade

Equipment
Vortex

Microfuge

Sterile 1.5 mL microfuge tubes, test tube racks (autoclaved or chemically disinfected)

Pipettor (10 μL, 200 μL, 1000 μL) and disposable filter (aerosol-resistant) tips

Cepheid Smartcycler II or equivalent thermoycler and 25 μL reaction tubes (Cepheid #900-0022), Cepheid reaction tube rack (Cepheid #900-0087), chilled to 4°C

Cepheid Microfuge

3.3.2 RT PCR Buffers and Enzymes

Qiagen One-step RT-PCR kit (Qiagen #210212)

a. Qiagen OneStep RT Enzyme Mix (Reverse transcriptase, Polymerase)
b. 5X Qiagen OneStep RT PCR Buffer
c. dNTP Mix
d. RNase-free water
e. RNasin (Promega #N211 or equivalent)
f. $MgCl_2$ 50mM (Invitrogen or equivalent)
g. 10mM forward primer, reverse primer, probe

3.3.3 96-Well Magnetic Bead for RNA Recovery (Manual Method)

MagMax-96 kit (*Ambion/Asuragen Inc.* #1835 or 1836-4 or equivalent).
Kit components are prepared according to the manufacturer's instructions.

Equipment

a. Magnet Stand (*Ambion/Asuragen* Inc, #10027 or 10050)
b. Vortex
c. 96-well Plate Shaker (*Lab-Line Instrument Inc* or equivalent).
d. Matrix Impact Equalizer 12-channel 250 μL and 1250 μL pipettors or equivalent.
e. Disposible pipet filter (aerosol-resistant) tips (250 μL, 1250 μL).
f. 96-well polystyrene round bottom plates for extraction steps (*Evergreen, Inc.* #222 8032-R1K or equivalent).
g. Foil plate sealer (Diversified Biotech #ALUM-1000). Foil sealed microtiter plates with mastermix can be stored for up to 7 days at –70°C.
h. Bio-Rad/Opticon thermocycler or equivalent for 96-well realtime PCR.
i. Hard-Shell® thin-wall 96-well skirted PCR plates, white shell, white well (Bio-Rad, #HSP-9655).
j. Optical flat 8-cap strips (Bio-Rad, #TCS-0803).

3.3.4 RT PCR Buffers and Enzymes

a. RNAse free water (Ambion/Asuragen Inc or equivalent)
b. Invitrogen RNA viral detection One-Step qRT-PCR UltraSense, #11732-101 (contains enzymes, Mg^{++}, dNTPs, buffer concentrate)
c. 100 mM forward primer, reverse primer, probe

3.5 Precautions

Beta-mercaptoethanol is toxic and should be dispensed in a chemical fume hood.

3.6 Procedure/Detail

Disinfect all equipment prior to starting the assay by gently wiping with 10% bleach solution, followed by wiping with 70% ethanol to remove any bleach residue.

3.7 Single Tube RNA Extraction and Realtime RT PCR

Prepare PCR Reagents and Buffers prior to handling clinical specimens and extracted nucleic acids.

Preparation of realtime RT PCR mastermix (buffers and enzymes),

1. Calculations for the volume of each reagent and enzyme needed are performed using a pre-programmed spreadsheet or manually-calculated worksheet.
2. Remove reagents and enzymes needed from $-20^{0°C}$ storage, quickly thaw, vortex to mix, pulse spin to remove any liquid adherent to the cap, and place the tubes on ice.
3. Working in a PCR-reagent clean biosafety cabinet, prepare mastermix by adding the appropriate volumes of each reagent as calculated for the number of samples to be tested (add 2 additional specimens to calculations to assure adequate final volume). Work in the same order as provided on the worksheet. *Note.The probe should always be the final reagent added.*

Avian Influenza virus Matrix gene Single-tube Worksheet						
Qiagen Enzyme Mix- Smartcycler qRT-PCR						
					Date	
Number of sample to be tested + 2	***Enter number***				**Batch #**	
					Initials	
Add in following order						
Master Mix	**Lot #**	**1 Rxn**	**Final Conc.**		**µL**	
Nuclease-free water		7.11			7.11	H_2O
5X Buffer		5	1X		5.00	5X Buffer
50 mM MgCl		0.63	3.75 µM		0.63	50 mM MgCl
AI matrix Forward Primer (10 µM)		1	0.4 µM		1.00	For Primer
AI matrix Reverse Primer (10 µM)		1	0.4 µM		1.00	Rev Primer
dNTP's		0.8	320 µM		0.80	dNTP's
RNasin 40 U/uL		0.17	6.65		0.17	Rnasin
Enzyme Mix		1			1.00	Enzyme Mix
AI matrix Probe (10 µM)		0.3	0.12 µM		0.30	Probe
		17	µL per Rxn		17.00	Total Volume

vNDV fusion gene - Single tube Worksheet				**Date**	
Invitrogen One-step - Smartcycler qRT PCR				**Batch #**	
				Initials	
Enter the number of samples	**1**				
(Add reagents in following order)					
Master mix	**Lot #**	**1 Rxn**	**Final Conc.**	**μL**	
Nuclease-free water		1		1.00	Water
5X Rxn Mix		4		4.00	5X Rxn Mix
vNDV-Forward primer (10 μM)		3	1.2 μM	3.00	vNDV-For
vNDV-Reverse primer (10 μM)		1.5	0.6 μM	1.50	vNDV-Rev
Enzyme Mix		1		1.00	Enzyme Mix
ND Virulence marker Probe (10 μM)		0.5	0.2 μM	0.50	Probe
		11	μL per Rxn	**11.00**	μL Total Volume

4. Vortex, then pulse spin (centrifuge) the mastermix tube. Wrap the tube in foil to protect from light and transfer to the area where the RNA will be added to the mastermix.

Note: To avoid contamination of PCR reagents it is critical that nucleic acids and clinical material not be handled in the PCR-reagent area, or that pipettors and equipment in this area never be used with clinical materials or nucleic acids.

3.8 RNA Extraction is Performed Per Kit Manufacturer's Instructions

Extract the clinical specimens, including the designated Positive Extraction Control and the Negative Extraction Control with each extraction run. For Quality Control purposes, it is critical that the extraction controls be handled and extracted at the same time and using the same reagents as the clinical specimens.

Total RNA testing should be initiated within 4 h following extraction. Within that time period, total RNA is stored at 4°C. Should the testing delay exceed 4 h, RNA is stored at –70°C or lower.

3.9 PCR Amplification and Detection

1. Place reaction tubes in the chilled tube rack, handling the tubes by the neck or lid rather than the sides to avoid contact with the "reading windows". Label the cap on the individual tubes to insure they are kept in the correct sequence.
2. For AI add 17 μL of mastermix to each reaction tube. For ND add 11 μL mastermix to each reaction tube.
3. For AI add 8 μL of each specimen RNA, closing the cap firmly before moving to the next tube. For ND add 14 μL of each specimen RNA. Include the positive extraction control, the negative extraction control. The positive PCR control and the negative PCR control with each assay (one or more specimens tested at the same time).
4. Centrifuge the reaction tubes using the Cepheid microfuge for 3–5 s.
5. Load the reaction tubes into the thermocycler; start the AI matrix run using

   50°C for 3 min
   95°C for 15min
   2 – temperature cycling for a total of 45 cycles
     94°C for 1 s
     60°C for 20 s

   Load the reaction tubes into the thermocycler; start the vNDV run using
     48°C for 30 min
     95°C for 2 min
   3 – temperature cycling for a total of 40 cycles
     94°C for 15 s
     58°C for 30 s
     72°C for 15 s

3.10 96-Well Magnetic Bead for RNA Recovery (Manual Method) and Realtime RT PCR

3.11 Prepare PCR Reagents and Buffers Prior to Handling Clinical Specimens and Extracted Nucleic Acids

3.12 Preparation of Realtime RT PCR Mastermix (Buffers and Enzymes)

1. Calculations for the volume of each reagent and enzyme needed are performed using a pre-programmed spreadsheet or manually-calculated worksheet,

2. Remove reagents and enzymes needed from –20°C storage, quickly thaw, vortex to mix, pulse spin to remove any liquid adherent to the cap, and place the tubes on ice.
3. Working in a PCR-reagent clean biosafety cabinet, prepare mastermix by adding the appropriate volumes of each reagent as calculated for the number of samples to be tested (plus +5 specimens or alternately 16% volume to assure adequate final volume). Work in the same order as provided on the worksheet. *Note.The probe should always be the final reagent added.*

AIV matrix gene 96-well						
Invitrogen ultrasense – opticon						
					Date made	
Enter the number of samples to be tested +5:		***Enter number***			Total made	
					Initials	
(Add reagents in this order)					Plate #	
Master mix	Lot #	1 Rxn	Final Conc.		µL	
RT-PCR Water		11.65			11.65	Water
5X Rxn Mix		4			4.00	5X Rxn Mix
Forward primer (100 µM)		0.15	0.6 µM		0.15	MA-For Primer
Reverse primer (100 µM)		0.15	0.6 µM		0.15	MA-Rev Primer
Enzyme mix		1			1.00	Enzyme Mix
Probe (100 µM)		0.05	0.2 µM		0.05	MA Probe
		17	µL per Rxn		**17.00**	**µL Total Volume**
1) Aliquot **17 µL** of Master Mix to each reaction tube.						
2) Add **8 µL** of target RNA.						

vNDV fusion protein gene 96-well Invitrogen UltraSense - Opticon					
				Date made	
Enter the number of samples:		**1**	*	Total made	
				Initials	
(Add reagents in this order)				Plate #	
Master mix	Lot #	1 Rxn	Final Conc.	**Ul**	
RT-PCR H2O		5.5		6.38	RT-PCR H2O
5X Rxn Mix		4		4.64	5X Rxn Mix
vNDV-Forward primer (100uM)		0.3	1.2 µL	0.35	vNDV-For
vNDV-Reverse primer (100uM)		0.15	0.6 µM	0.17	vNDV-Rev
Enzyme Mix		1		1.16	Enzyme Mix
Virulence marker Probe (100uM)		0.05	0.2 µM	0.06	VFP Probe
		11	µL per Rxn	**12.76**	µL Total Volume
*includes 16% additional volume					
a) Aliquot 11 µL of Master Mix to each reaction tube.					
b) Add 14 µL of target RNA.					

4. Pipette master mix into the Bio-Rad white 96-well plate. Using a Matrix pipettor, load 250 µL mastermix and dispense the assay-specific µL volume into each well.
5. Cover plates with foil plate sealers to protect from light and evaporation; sealed plates with mastermix can be stored for up to 7 days at –70°C.

3.13 RNA Extraction Is Performed Per Kit Manufacturer's Instructions

Extract the clinical specimens, including the designated Positive Extraction Control and the Negative Extraction Control with each extraction run. For Quality Control purposes, it is critical that the extraction controls be handled and extracted at the same time and using the same reagents as the clinical specimens.

Total RNA testing should be initiated within 4 h following extraction. Within that time-frame, total RNA is stored at 4°C. Should storage exceed 4 h, RNA is stored at –70°C or lower.

3.14 Specimen Preparation

Specimens should be in a standardized tube size to facilitate multi-channel transfer, example 2 mL vials or tubes (Sarsted #72.694.006) or equivalent.

1. Vortex samples gently prior to RNA extraction. Do not centrifuge.
2. Working with 1 row (12 specimens) at a time, uncap and place tubes in a separate rack dedicated to the extraction process.
   a. Transfer 50 μL of each row of specimens to the corresponding row of the 96-well plate.
   b. Between rows, replace the cover on the microtiterplate.
   c. Cap the tubes and transfer to the storage box before moving to the next row of samples.

   Note. Because multiple specimens are handled in sequence, it is good lab practice to check gloved fingertips for any evidence that fluid from the caps or tubes has not contaminated your gloves. If any moisture is dedicated, decontaminate with 10% bleach and dispose of gloves. Wipe the tubes, rack, and immediate work area with 10% bleach and again with ethanol before continuing.

3. Quality Control specimens are located in wells interspersed with diagnostic specimens to appropriately control for cross-contamination during extraction and amplification steps, e.g. row C well 3, row F well 8.
   a. Pipette 50 μL of negative extraction control into the designated well.
   b. Dispose of pipette tip in bleach solution.

c. Pipette 50 μL of positive extraction control (LPAI or armored RNA UCD#3) into two designated wells. Dispose of tip in bleach solution.
d. QC specimen vials are disinfected by wiping with 70% alcohol after each use.
e. Leave wells H11 and H12 empty for PCR positive and negative controls, which will be added directly to the wells containing PCR mastermix immediately prior to qRT PCR amplification.

4. Use care to keep the microtiter plate flat. Store the microtiter plate containing specimens at –20°C if not proceeding immediately to RNA extraction steps.

3.15 96-Well RNA Extraction General Recommendations

- To remove fluid, program Matrix pipettor to slow speed, add fluid to the 96-well plate with pipettor programmed on medium speed.
- To remove fluid from the 96-well plates during the extraction and wash steps, leave the plate on the magnetic stand; To add extraction and wash fluids to the wells, remove the plate from the magnetic stand.
- Magnetic beads used in the extraction process should be suspended after adding the elution solution, pipette gently if the beads are clumped or not suspended.

3.16 Total RNA Isolation

Preparation. Place bead solution on vortex and shake on setting 3 for 2 min.

1. If samples have been pre-plated and the 96-well plate frozen, thaw in a biosafety cabinet with the plate covers off to prevent condensation from running onto plate surface.
2. Double check the 96-well plate for loading errors (empty wells, excessive volume in wells), and for missing plate numbers/missing bar codes.

3.17 RNA Binding Using a Programable Matrix or Equivalent Pipettor

1. Add 101 μL Lysis/Binding solution into each well, and shake for 5 min.
   Shaker dial position, Medium speed *Fill 1250 μL/dispense 101 μL*

2. Add 20 μL beads/binding mix in each well of a U-bottom plate, and shake for 4 min.
   Shaker dial position, Medium speed *Fill 250 μL/dispense 20 μL*
3. Move the U-bottom plate onto the Magnetic Stand, wait 2 min for beads to pellet. Slowly remove supernatant. *Fill 240 μL/dispense 240 μL*
4. Add 100 μL prepared Wash Solution I, shake for 1 min (it is possible that the beads won't break up, there is not a negative effect on RNA quality or yield).
   Shaker dial position, Medium speed+ *Fill 1250 μL/dispense 100 μL*
5. Move the U-bottom plate on to the Magnetic Stand, wait 1 min to pellet beads. Remove supernatant. *Fill 240 μL/dispense 240 μL*
6. Add 100 μL prepared Wash Solution II, shake for 30 s.
   Shaker dial position, Medium speed+ *Fill 1250 μL/dispense 100 μL*
7. Move the U-bottom plate on to the Magnetic Stand, wait 30 s to pellet beads. Remove supernatant. *Fill 240 μL/dispense 240ul*
8. Add 100 μL prepared Wash Solution II, shake for 30 s.
   Shaker dial position, Medium speed+ *Fill 1250 μL/dispense 100 μL*
9. Move the U-bottom plate on to the Magnetic Stand, wait 30 s to pellet beads. Slowly remove supernatant. Shake plate for 2 min dry to help the evaporation of ETOH (part of the washing solution). *Shaker dial position, High speed*

RNA Elution

10. Add 30 μL Elution Solution to each well, shake for 4 min. Beads should be suspended, otherwise use a manual 200 μL pipettor to re-suspend the beads by pipetting up and down.
    Shaker dial, High speed *Fill 250 μL/dispense 30 μL*
11. Transfer of RNA to a 96-well plate containing the assay-specific qRT-PCR mastermix. If the RNA will not be used for PCR amplification within 4 h, transfer the RNA into a clean plate (place the U-bottom plate on the Magnetic Stand, wait 1 min to pellet the beads. Transfer the supernatant into a clean plate (*25 μL per well*), cover with an adhesive plate sealer, wrap with aluminum foil and freeze the plate at –70°C.)

    Note. Manual 12 and 8 channel pipettors are not as efficient for the extraction based on slower and less forcefuldispensing of fluid, which is needed to help disperse the magnetic beads.

3.18 PCR Amplification and Detection

1. To perform the Avian Influenza matrix gene RT-PCR, 8 μL of extracted RNA is pipetted into each well of the AI matrix mastermix plate containing 17 μL per well of master mix using the Matrix Equalizer pipettors.
2. Positive and Negative PCR control RNA (8ul) is added respectively to the last two wells of each microtiter plate.
3. Load the reaction plates into the thermocycler; start the run using
   48°C for 30 min
   95°C for 2 min
   3-step cycling for a total of 45 cycles
      94°C for 15 s
      60°C for 30 s
      72°C for 15 s
4. To perform the vNDV fusion protein gene RT-PCR, 11 μL of extracted RNA is pipetted into each well of the AI matrix mastermix plate containing 17 μL per well of master mix using the Matrix Equalizer pipettors.
5. Positive and Negative PCR control RNA (8ul) is added respectively to the last two wells of each microtiter plate.
6. Load the reaction plates into the thermocycler; start the run using
   48°C for 30 min
   95°C for 2 min
   3-step cycling for a total of 45 cycles
      94°C for 15 s
      60°C for 30 s
      72°C for 15 s

3.19 Expected Values

In order for the assay to be valid

1. The extraction and PCR negative control must have no Ct crossing the threshold.
2. The Virus extraction and PCR positive controls must give a Ct within 2 standard deviations of the expected value (see example, AI qRT PCR).The target Ct for optimum assay precision is the expected value ± 1 Ct for the controls. Validated Armored RNA controls can replace positive extraction controls.
3. The difference between the extraction control and PCR positive controls must be within 3 Ct of the expected value. Optimum performance is within 1.5 Ct of the expected value.
4. The difference between the two extraction controls must be less than 2 Ct.

Example, values are established for each reference control lot produced*

Expected value	*Mean Ct*	*Ct Std dev*	*Target Ct (± 1 Ct)*	*Acceptable Ct (± 2 SD)*
AI extraction	*18.2*	*1.6*	*17.2–19.2*	*15–21.4*
AI PCR	*17.6*	*1.4*	*16.6–18.6*	*14.8–20.4*
Difference	*0.6*	*0.15*	*≤1.6*	*0.3–0.9*

*Expected Ct value for new lots of extraction and paired PCR controls are based on triplicate side-by-side testing with established controls. Current values are recorded in the QC form.

The linear range of the influenza matrix gene assay is between 12 and 35 cycles.

A specimen is identified as positive for target sequence when a positive Ct is detected, and the resulting realtime PCR fluorogram shows a log-linear increase in fluorescence. Post-analytic procedures are used for Ct >35 and where the fluorogram does not demonstrate a typical log-linear increase in fluorescence.

3.20 Post Analytical Procedures

The realtime PCR amplification procedure may be followed by 4% agarose gel electrophoresis for confirmation of amplicon size. Confirmation of amplicon size is not required, but is an additional practice used for RNA viruses where sequence variability and mutability of the RNA genome may result in changes in assay performance over time or geographic region. Gel electrophoresis verification of amplicon is recommended for all specimens with a Ct of 35 or greater cycles, to aid in differentiating weak positive realtime PCR results from late cycle probe decay.

3.21 Bibliography and Related Documentation

96-well extraction and PCR

1. http://www.ambion.com/techlib/posters/thruput_mag_0402.htmL (03-2005)
2. http://www.ambion.com/techlib/prot/fm_1830.pdf (03-2005)
3. Crossley et al., High-throughput realtime RT PCR assay to detect exotic Newcastle disease virus during the California 2002–2003 outbreak

3.22 Primer and Probe Design and Performance

4. Spackman et al. 2002. Development of a real-time reverse transcriptase PCR assay for type A influenza virus and the avian H5 and H7 hemagglutinin subtypes. *J. Clin. Microbiol.*, **40** (9) 3256–60.
5. Spackman et al. 2008. Detection and identification of the H5 hemagglutinin subtype by real-time RT-PCR. *Methods Mol. Biol.*, **436**, 27–33.
6. Spackman et al. 2008. Type A influenza virus detection and quantitation by real-time RT-PCR. *Methods Mol. Biol.*, **436**, 19–26.
7. Aldous EW., et al. 2001 Rapid pathotyping of Newcastle disease virus (NDV) using fluorogenic probes in a PCR assay. *Vet Microbiol.*, **80** (3), 201–12.
8. Creelan et al. 2002. Detection and differentiation of pathogenicity of avian paramyxovirus serotype 1 from field cases using one-step reverse transcriptase-polymerase chain reaction. *Avian Patho.*, **31** (5), 493–9.

4 Guidance Document- Assay Failure Response

4.1 Assay Performance

Numerous factors contribute to the reliability, repeatability, and reproducibility of molecular-based diagnostic tools, including but not limited to;

1. Environmental conditions
2. Variation in acceptable ranges of performance for different pieces of equipment or equipment platforms
3. Measurement traceability
4. Pre-test sampling and handling of specimens
5. Reagent performance variability
6. Human error

It is diagnostic service policy to track QC sample performance over time and establish expected performance ranges (e.g. 95% confidence interval, standard deviation from the mean, % coefficient of variation). Once established, the performance expectations are monitored and recorded for *every* assay performed where diagnostic specimens are included and results reported.

4.2 Assay Failure

If one or more assay controls fail on a assay or assay run; the technician will consult with the section supervisor or designee as to further action. Regardless of follow-up approach used, all assay runs are recorded, and values for failed QC samples are documented and tracked. Where an explanation for the failure is possible, that is additionally documented and linked to the QC sample values obtained for the failed run.

The most typical response to an initial failure is to repeat the assay or assay run, including controls and diagnostic specimens, recognizing that biologic variability and human error are the most likely source of assay failure.

If assay controls fail on sequential assays (at least 2 sequential runs), or a specific assay has a pattern of >10% failure over time, the technical staff will initiate an assay failure investigation including documentation, which may include but is not limited to:

1. Alternate vial or lot of assay control(s)
2. Alternate technician
3. Alternate vial or lot of specific reagent(s) (e.g. primer, probe, DNase, and etc.)
4. Alternate but equivalent extraction method
5. Alternate but equivalent piece of equipment or equipment platform (e.g. pipettor, thermocycler, etc.)
6. Alternate aliquot of specimen if available

4.3 Performance Troubleshooting Guidelines

- 4 or more consecutive control values exceed the mean ± 1 standard deviation. Performance is within the acceptable range, but the trend is often an indicator of need for equipment maintenance or reagent decay.
- 2 or more consecutive control values exceed the mean ± 2 Standard deviations suggests systematic error.
- Any control value exceeds 3 standard deviations within an individual run suggests random error.
- 10 consecutive control values are on the same side of the mean serves as an indicator of need for equipment maintenance or reagent re-calibration (e.g. stock reagent dilution error).

SOP 16. Preparation of Silica Particles for Nucleic Acid Extraction

LAB-ON-SITE
EU PROJECT
SSPE-CT-2004-51364

Courtsey of Thomas Bruun Rasmussen
National Veterinary Institute, Technical University of Denmark,
Department of Virology,
Lindholm DK-4771 Kalvehave
Denmark

Responsible scientist Thomas Bruun Rasmussen
E-mail thb@vet.dtu.dk

Created 2008-02-04
Updated 2008-02-04

1 Introduction

This SOP describes the preparation of Silica particles for the extraction of nucleic acids. DNA and RNA bind efficiently to acid-washed Silica particles [1]. This procedure rescues the nucleic acids from the main part of debris and interfering agents in serum and organ materials. The current method is used at DTU Vet [2]

1.1 Principle

Silica particles are sedimented twice in water, to remove the particles with a suboptimal size. Finally HCl is added to the suspension.

1.2 Safety

This procedure involves 32% HCl
Risks involved, C; R-34, R-37

1.2.1 Virus Safety

No virus is used in this procedure, however, it is essential to perform the procedure in a clean area.

1.3 Equipment

Glass cylinder, 27.5 × 5 cm
Pipettes
Wide bottomed polyethylene-bottle for storage of stock
Eppendorf tubes for aliquots
Parafilm
Disposable gloves

1.4 Reagents

SiO_2 (Sigma S-5631).
Milli-Q water.
32% HCl

1.5 Waste Treatment

All waste from this procedure can be disposed of without further precautions.

2 Procedure

a. Weigh 60 g of Silica particles and add to a cylindrical glass tube (27.6 × 5 cm).
b. Add Milli-Q water to a total volume of 500 mL.
c. Seal with parafilm and turn the glass tube several times until Silica particles are in suspension.
d. Leave to sediment 24 h at RT.
e. Pour carefully (or use suction) 440 mL supernatant off.
f. Resuspend by adding Milli-Q water to 500 mL, mix as in “3”.
g. Leave to sediment for 5 h at RT.
h. Use suction to remove 440 mL supernatant. Leave 60 mL of residual volume, this will contain aprox. 50 vol% Silica.
i. Shake well and add 600 μL 32% HCl.

j. Keep the stock in a high density polyethylene or glass bottle in the dark at RT, sealed with parafilm to reduce evaporation. This stock will keep for years.
k. Make 200–500 µL aliquots in Eppendorf tubes for everyday use. Mark date of stock production.
l. Vortex thoroughly before use. Is it essential that the Silica is in suspension before use.

3 Bibliography

1. Boom, R, Sol, CJ, Salimans, MM, Jansen, CL, Wertheim-van Dillen, PM, van der Noordaa, J. 1990. Rapid and simple method for purification of nucleic acids. *J. Clin. Microbio.*, **28**, 495–503.
2. Oleksiewicz, MB, Botner, A, Madsen, KG, Storgaard, T. 1998. Sensitive detection and typing of porcine reproductive and respiratory syndrome virus by RT-PCR amplification of whole viral genes. *Vet. Microbiol.*, **64**, 7–22.

SOP 17. Boom-Silica RNA Extraction (GuSCN, phenol, Silica)

LAB-ON-SITE
EU PROJECT
SSPE-CT-2004-513645

Courtesy of Thomas Bruun Rasmussen
National Veterinary Institute, Technical University of Denmark,
Department of Virology,
Lindholm DK-4771 Kalvehave
Denmark

Responsible scientist
Thomas Bruun Rasmussen E-mail thb@vet.dtu.dk

Created 2008-02-04
Updated 2008-02-04

1 Introduction

This SOP describes extraction of nucleic acids from serum or organ material for subsequent RT-PCR analysis. The content of PCR-interfering agents is reduced thus resulting in an increased sensitivity of the PCR. Detection of RNA viruses by PCR is often hampered by the interference of organic materials on the efficacy of PCR (Konet et al., 2000).

2 Definition

RT-PCR – reverse transcriptase-polymerase chain reaction
GuSCN – guanidine isothiocyanate
RT – room temperature
μL – microliter

3 Principle

This procedure combines several steps each reducing the interference of organic materials on the efficacy of PCR. Initially the material is treated with guanidine isothiocyanate (GuSCN) to avoid RNAse activity and to inactivate virus in the sample. Organ materials are extracted with acid phenol (for RNA viruses, here classical swine fever) and 1-Bromo-3-chloropropan. The nucleic acids are subsequently isolated by binding to acid-washed silica particles [1]. Following washes in GuSCN/citrate and ethanol, nucleic acids are eluted in RNAse free water and are now ready for the RT-PCR procedure.

4 Safety

All procedures in this description must be performed in a flow-hood.

GuSCN , Xn; R20/21/22, R 32
Phenol , Tx, C; R24/25, R34
1-Bromo-3-chloropropan (C3H6BrCl), Xn; R 22, R 40

4.1 Virus Safety

Any virus will be inactivated by the high molar GuSCN used. However, to reduce the risk of contamination the procedure is performed in a clean area, i.e. PCR lab outside (mastermix extraction and preparation area). Changing of clothes and showering before entering the lab is required. All waste materials are incinerated immediately on the island.

5 Equipment

Safety laminar air flow hood (must be approved for work with
1-Bromo-3-chloropropan and phenol)
Disposable gloves
Pipettes
Plugged tips only
Positive displacement pipette for 1-Bromo-3-chloropropan
Eppendorf tubes (1.5 and 2 mL)
PCR tube rack
Mini centrifuge
Rotating spear
Heating block

6 Reagents

6.1 5 M GuSCN (produced at the Pharmacy of the Royal Vet University)
GuSCN (Fluka 50990) 591 g
Distilled water up to 1 L
Keep in a dark bottle, at RT.

6.2 1 M Citric acid pH 5,2 (produced at the Pharmacy of the Royal Vet University)
Citric acid monohydrate (Merck 1.00244) 4.90 g.
Sodiumcitrate dihydrate (Merck 1.06448) 23.52 g.
Sterile water up to 100 mL.
Autoclave and store at +4°C.

6.3 GuSCN/citrate (working solution of GuSCN)
30 mL 5 M GuSCN + 936 μL 1 M citric acid, pH 5.2
Store at +4°C, shelf life 1 month.

6.4 Elution buffer
600 μL nucleasefree water + 20 μL PRIME RNase inhibitor.
70% EtOH
31.5 mL 99.9% EtOH + 13.5 mL Milli-Q water

6.5 2 M NaOAc, pH 4.2, biotechnology grade (Amresco E 502), store at RT

6.6 Phenol pH 4.3 (Amresco), store at +4 °C

6.7 1-Bromo-3-chloropropan 24 parts + 1 part isoamylalcohol (Amresco X 205), store at RT

6.8 Nuclease free water (Amresco), stored at –20°C in aliquots of 1 mL.

6.9 PRIME RNAse inhibitor (5 prime – 3 prime, Inc., Boulder, (CO), USA)

6.10 99.9% EtOH (Danisco distillers, Aalborg, Denmark)

6.11 Milli-Q water.

7 Procedure

7.1 Pretreatment of Samples

Add 96 µL serum to 304 µL GuSCN/citrate, vortex well.

At this stage the material may be stored at –20°C.

If the material causes debris (organ suspensions, semen or full blood) centrifuge and transfer supernatant to clean tube.

7.2 Phenol/1-Bromo-3-chloropropan Extraction of Samples

Steps 2–4 are only relevant for organ materials

Thaw the samples, vortex and spin down, each tube is treated as follows,

Add 40 µL of NaOAc.

Add 400 µL of cold acid phenol; mix by turning the tubes several times.

Add 120 µL 1-Bromo-3-chloropropan, use positive displacement pipette. Shake the tubes vigorously to mix the material.

Spin the tubes at 11,300 g for 2 min.

Transfer the supernatant (400 µL) to clean 1.5 mL Eppendorf tubes.

The tubes may be frozen at this step or go immediately to Silica binding.

7.3 Binding of RNA to Silica Particles

Add 2.5 µL Silica [AI-1.01.602] to each pretreated sample.

Vortex briefly and leave the tubes for 20 min at RT on a rotating spear.

Prepare washing solution, set the heating block for 65°C, take 1 mL nuclease free Amresco water from the freezer.

Spin at 2400 ×*g* for 20 s, remove the supernatant carefully using a pipette.

Wash Silica twice using 475 µL GuSCN/citrate (vortex, spin 20 s, 2400 g, and remove the supernatant by a pipette). It is essential that the Silica is 100% resuspended after each pelleting.

Wash Silica twice using 475 µL 70% EtOH (vortex, spin 20 s, 2400 g, and remove the supernatant by a pipette).

Wash Silica once using 475 µL 99.9% EtOH (vortex, spin 20 s, 2400 g, and remove the supernatant by a pipette).

Spin the dried sample again (2400 g for 10 s) and remove as much supernatant as possible.

Let the Silica dry for 5–10 min in the flow hood.

Resuspend the pellet in 22.5 μL elution-buffer, vortex thoroughly to completely resuspend the Silica.

Incubate the tube for 5 min at 65°C in the heating block.

Spin tubes at 11,300 g for 2 min, transfer the supernatant to new Eppendorf tubes, the RNA samples are now ready for cDNA synthesis. Discard of the tubes containing pellets.

8 Bibliography

1. Boom, R, Sol, CJ, Salimans, MM, Jansen, CL, Wertheim-van Dillen, PM, van der Noordaa, J. 1990. Rapid and simple method for purification of nucleic acids. *J. Clin. Microbiol.*, **28**, 495–503.
2. Konet, DS, Mezencio, JM, Babcock, G, Brown, F. 2000. Inhibitors of RT-PCR in serum. *J. Virol. Methods*, **84**, 95–8.
3. Uttenthal, A, Storgaard, T, Oleksiewicz, MB, de Stricker, K. 2003. Experimental infection with the Paderborn isolate of classical swine fever virus in 10-week-old pigs, determination of viral replication kinetics by quantitative RT-PCR, virus isolation and antigen ELISA. *Vet. Microbiol.*, **92** (3), 197–212

SOP 18. Mab Based Competitive ELISAs for H5 and H7 Antibody Detection in Avian Sera

LAB-ON-SITE
EU PROJECT
SSPE-CT-2004-51364

Courtesy of Paolo Cordioli
Istituto Zooprofilattico Sperimentale della Lombardia e dell'Emilia Romagna
Reparto Biotecnologie – Centro Nazionale di Referenza per le Malattie Vescicolari
Via Bianchi, 9
25124 Brescia
Italy

Responsible scientist Paolo Cordioli
E-mail paolo.cordioli@bs.izs.it

Created 2008-02-13
Updated 2008-02-15

1 Introduction

This SOP describes two ELISA assays to detect antibodies against H5 and H7 avian influenza viruses (AIV) in avian sera. These tests were developed by Partner 7, IZSLER, Brescia. The principle of the two assays is the same; it is a competitive ELISA that uses the H5 and H7 partially purified AIVs coated to the plate and the conjugated Mabs specific for H5 (5D8) and H7 (7A4) respectively. The ability of test sera to inhibit the binding of peroxidase-conjugated MAbs to the antigen is then evaluated.

2 Reagents

2.1 ELISA Buffers

2.1.1 Coating Buffer (0.05 M Carbonate-Bicarbonate Buffer, pH 9.6)

Na_2CO_3	1.59 g
$NaHCO_3$	2.93 g
Distilled water to	1000 mL
Store at +4°C	

2.1.2 Phosphate Buffered Saline (PBS), pH 7.4

	Normal strength	×10 strength
NaCl	8.0 g	80 g
KH_2PO_4	0.2 g	20 g
$Na_2HPO_4.12H_2O$	2.9 g	29 g
KCl	0.2 g	2g
Distilled water to Store at RT	1000 mL	1000 mL

2.1.3 Washing Buffer (PBS-Tween)

PBS (normal strength) + Tween 20 (PI379 Sigma) 0.05%.
Store at RT.

2.1.4 Sera and Conjugate Diluent Buffer

PBS-Tween 20 + yeast extract 1%
Prepare the minimum needed volume every day and use it fresh.

2.1.5 Chromophore/Substrate Solution

OPD (orthophenylendiamine- Sigma) 0.5 mg/mL in phosphate-citrate buffer
Phosphate-citrate buffer, pH 5

Citric Acid 0.1 M	98.6 mL
Na_2HPO_4 0.2 M	101.4 mL
Store at 4°C	

Dissolve tablets of OPD (SIGMA) in the appropriate volume of buffer (e.g. one tablet of 60 mg OPD in 120 mL of phosphate-cytrate buffer, pH 5). This solution can be stored at –20°C in appropriate aliquots (6, 12, 18 or 24 mL aliquots, suited respectively for 1, 2, 3 or 4 plates) in the dark. Immediately before use add H_2O_2 at 0.02% final concentration (e.g. H_2O_2 30% 4 µL H_2O_2 per 6 mL OPD) potential carcinogenic agent R45.

2.1.6 Blocking Solution (H_2SO_4 2N)

H_2SO_4 96%	56 mL
Distilled water	944 mL

Store at room temperature.

3 Biological Reagents

3.1 Antigens

3.1.1 H5 AIV

Allantoic fluid from SPF chicken embryonated eggs infected with the reference strain A/Tk/En/28/73 H5N2 LPAI.

3.1.2 H7 AIV

Allantoic fluid from SPF chicken embryonated eggs infected with the field strain A/Tk/It/2676/99 H7N1 LPAI.

3.1.3 Antigen Production

Viruses are inoculated into 9–11-day-old embryonated SPF eggs. After 3 days of incubation at 37°C, allantoic fluids (AF) are collected and then inactivated with ß-propiolactone. Viruses are concentrated and partially purified by centrifuging the AF at low speed (3000g for 20min), then layering it on 25% (w/w) sucrose gradient and pelleting the virus by centrifugation at 100.000g at 4°C for 2 h.

The pellet is re-suspended in PBS (100X) and finally mixed with glycerol 50% and stored at –20°C. Optimal antigen dilution for ELISA is pre-determined by examined a two-fold dilution series with the appropriate conjugated Mab.

3.2 Positive Control Sera

3.2.1 Positive H5 Control Serum

Sera from SPF chickens experimentally infected with A/Tk/It/90302/05 H5N2 LPAI.

3.2.2 Positive H7 Control Serum

Sera from SPF chickens experimentally infected with A/Tk/It/2676/99 H7N1 LPAI.

3.2.3 Negative Control Serum

SPF chicken sera.

3.3 Conjugated Mab

Peroxidase-conjugated Mabs anti-H5 (5D8) and anti-H7 (7A4).

They can be mixed with glycerol 50% and stored at –20°C or pre-diluted in stabilizing buffer and stored at +4°C.

Prepare working conjugate dilution immediately before use.

Mabs 5D8 and 7A4 were produced by IZSLER.

4 Assay Procedure

4.1 Antigen Coating

Plates are coated with 50 μL/well of the H5 or H7 antigen respectively, optimally diluted in carbonate buffer pH 9.6, and incubated overnight at 4°C.

4.2 Washing

Wash the ELISA plates three times (3 min/wash) by filling wells with 150–200 μL/well of washing Buffer at appropriate dilution.

4.3 Sera

Distribute 50 μL/well of test sera and control sera, diluted in PBS-Tween containing 1% yeast extract. 42 test sera (2 duplications/serum) can be examined in one plate, they are usually distributed and diluted directly in antigen coated wells, according the enclosed scheme (Fig. 5.1).

- Deliver 80 μL of diluent buffer in wells of the rows A, C, E, G (excluding wells A12, C12, E1, G1) and 50 μL in wells A12, C12, E1, G1
- Add 20 μL of test sera in wells of the rows A, C, E, G (excluding wells A12, B12, C12, D12, E1, F1, G1, H1, and G11, G12).
- Add 20 μL of each positive and negative control sera, following the scheme in Fig. 5.1, in wells G11 G12
- Mix using either a multichannel pipette or a shaker, obtaining the dilution 1/5
- Using a multichannel pipette tranfer 50 μL from wells of rows A, C, E, G (excluding wells A12, B12, C12, D12, E1, F1, G1, H1) to the corresponding wells of rows B, D, F, H to duplicate the dilution 1/5.

Serum dilution		**1**	**2**	**3**	**4**	**5**	**6**	**7**	**8**	**9**	**10**	**11**	**12**
1/5	**A**	1	2	3	4	5	6	7	8	9	10	11	100%
1/5	**B**	1	2	3	4	5	6	7	8	9	10	11	100%
1/5	**C**	12	13	14	15	16	17	18	19	20	21	22	100%
1/5	**D**	12	13	14	15	16	17	18	19	20	21	22	100%
1/5	**E**	100%	23	24	25	26	27	28	29	30	31	32	33
1/5	**F**	100%	23	24	25	26	27	28	29	30	31	32	33
1/5	**G**	100%	34	35	36	37	38	39	40	41	42	C+	C-
1/5	**H**	100%	34	35	36	37	38	39	40	41	42	C+	C-

Fig. 5.1 Micro plate layout (scheme of sera distribution)

1–42: Test sera; dilution 1/5, two duplications/serum
C neg Negative control serum, dilution 1/5, two duplications/serum
C pos H5 or H7 positive control serum, dilution 1/5, two duplications/serum
100% Reaction control wells (0% inhibition)

4.4 Distribution of Conjugates Mab, Without Washing

Add 25 µL /well of peroxidase-conjugated Mabs H5 or H7, diluted in PBS-Tween with 1% yeast extract.

Cover and incubate plates for 1 h at 37°C (possibly on a rotary shaker).

4.5 Washing

As above

4.6 Substrate

The colorimetric reaction is developed by distributing 50 µL/well of the substrate solution (OPD 0.5 mg/mL in phosphate-citrate buffer pH 5, 0.02% H_2O_2); after 10 min. at room temperature, the reaction is stopped by adding 50 µL/well of H_2SO_4 2N.

4.7 Reading

Absorbance values are read at 492 nm wave-length, using a micro plate reader.

5 Evaluation of Results

The pre-determined optimal dilutions of peroxidase-conjugated Mabs are expected to give spectrophotometric readings included in the range 1.5± 0.5 absorbance units (average of 8 wells 100% reaction = 0% inhibition) in control wells

Percentage inhibition produced by test and reference sera are calculated as follows,

$$\% \text{ inhibition} = 100 - (\text{serum OD} / 100\% \text{ OD}) \times 100$$

The negative control serum usually inhibits less than 50%

The positive control serum is expected to give ≥ 75% inhibition

Test sera are considered

Positive when the arithmetic mean of the two replicates produces an inhibition ≥ 75%

Negative when arithmetic mean of the two replicates produces an inhibition ≤ 75%.

SOP 19. Type A, H5, and H7 Avian Influenza Antigen Detection ELISAs

LAB-ON-SITE
EU PROJECT
SSPE-CT-2004-513645

Courtesy of Paolo Cordioli
Istituto Zooprofilattico Sperimentale della Lombardia e dell'Emilia Romagna
Reparto Biotecnologie – Centro Nazionale di Referenza per le Malattie Vescicolari
Via Bianchi, 9
25124 Brescia
Italy

Responsible scientist Paolo Cordioli
E-mail paolo.cordioli@bs.izs.it

Created 2008-02-13
Updated 2008-02-15

1 Introduction

This SOP describes three ELISA assays to identify type A, H5 and H7 avian influenza viruses using monoclonal antibodies (Mabs) specific for the nucleoprotein A (NPA), and H5 and H7 haemoagglutinin antigens, respectively. These tests were developed by Partner 7, IZSLER, Brescia. The principle of the three ELISAs is the same and is based on the use of capture Mabs specific for NPA, H5, and H7 AIVs respectively, coated onto 96 wells ELISA microplates, the addition of samples from allantoic fluids of infected embryonated eggs, and, finally, the addition of the same capture Mabs labelled with HRP. The wells where AIV antigen has been captured and then reacted with the specific conjugate are identified by mean of a OPD staining.

2 Reagents

2.1 ELISA Buffers

2.1.1 Coating Buffer (0.05 M Carbonate-Bicarbonate Buffer, pH 9.6)

Na_2CO_3	1.59 g
$NaHCO_3$	2.93 g
Distilled water to	1000 mL
Store at +4°C	

2.1.2 Phosphate Buffered Saline (PBS), pH 7.4

	Normal strength	X10 strength
NaCl	8.0 g	80 g
KH_2PO_4	0.2 g	20 g
$Na_2HPO_4 12H_2O$	2.9 g	29 g
KCl	0.2 g	2 g
distilled water to store at room temperature.	1000 mL	1000 mL

2.1.3 Washing Buffer (PBS-Tween)

PBS (normal strength) + Tween 20 (PI379 Sigma) 0.05%.
store at room temperature.

2.1.4 Sample and Conjugate Diluent Buffer

PBS-Tween 20 + yeast extract 1%
Prepare the minimum needed volume every day and use it fresh.

2.1.5 Chromophore/Substrate Solution

OPD, orthophenylendiamine (Sigma) 0.5 mg/mL in phosphate-citrate buffer
Phosphate-cytrate buffer, pH 5

Citric Acid 0.1 M	98.6 mL
Na_2HPO_4 0.2 M	101.4 mL
Store at 4°C	

Dissolve tablets of OPD (SIGMA) in the appropriate volume of buffer (e.g. one tablet of 60 mg OPD in 120 mL of phosphate-citrate buffer, pH 5). This solution can be stored at –20°C in appropriate aliquots (6, 12, 18 or 24 mL aliquots, suited respectively for 1, 2, 3 or 4 plates) in the dark. Immediately before use add H_2O_2 at 0.02% final concentration (e.g. H_2O_2 30% 4 µL H_2O_2 per 6 mL OPD) potential carcinogenic agent R45.

2.1.6 Blocking Solution (2N- H_2SO_4)

H_2SO_4 96%	56 mL
Distilled water	944 mL

Store at RT.

3 Biological Reagents

3.1 Catching Mabs

The solution of the catching Mabs (Mab anti-NPA, 5F10, Mab anti-H5, 5D8, and Mab anti-H7, 7A4) is a semi-purified concentrated Mab prepared in PBS containing glycerol 50% and stored at –20°C. Final dilution range 2–5 µg/mL depending on MAb.

Prepare working dilution immediately before use.
Mabs 5F10, 5D8 and 7A4 were produced by IZSLER.

3.2 Conjugated Mabs

Peroxidase-conjugated Mabs anti-NPA (5F10), anti-H5 (5D8) and anti-H7 (7A4). They can be mixed with glycerol 50% and stored at –20°C or pre-diluted in stabilizing buffer and stored at +4°C.

Prepare working conjugate dilution immediately before use.
Mabs 5D8 and 7A4 were produced by IZSLER.

3.3 Positive Controls

Positive AIV control sample
H1N1 infected allantoic fluid inactivated with ß-propiolactone (BPL).

Positive H5 control sample
H5N2 infected allantoic fluid inactivated with BPL.

Positive H7 control sample
H7N1 infected allantoic fluid inactivated with BPL.

3.4 Negative Control Sample, SPF Allantoic Fluid

Positive and negative control samples should be stored at –20°C in appropriate aliquots (200µL is suitable for one plate).

4 Assay Procedure

4.1 Coating

Nunc 96 wells microplates are coated overnight at 4°C with the capture Mabs anti NPA, anti-H5 and anti-H7 respectively, optimally diluted in coating buffer.

4.2 Washing

Washing the ELISA plates three times (3 min/wash) by filling wells with 150–200 µL/well of washing buffer at appropriate dilution.

4.3 Samples

Addition of 50 µL/well of allantoic fluids. Samples have to be tested in duplicate and it is recommended to test two dilutions (undiluted and ½) for each sample. For each plate, a AIV or H5 or H7 positive control and negative control have to be included.

Incubate plates for 1 h at 37°C.

4.4 Washing

Wash plates as described above.

4.5 Conjugated Mabs

Addition of 50 µL/well of the appropriate dilution of NPA or H5 or H7 specific conjugate respectively. Incubate for 1 h at 37°C.

4.6 Washing

As above.

4.7 Substrate

The colorimetric reaction is developed by distributing 50 µL/well of the substrate solution (OPD 0.5 mg/mL in phosphate-citrate buffer pH 5, 0.02% H_2O_2); after 10 min. at room temperature, the reaction is stopped by adding 50 µL/well of H_2SO_4 2N.

4.8 Reading

Absorbance values are read at 492 nm.

4.9 Evaluation of Results

The reactivity of AIV, H5 and H7 control antigen is expected to be higher than 0.2 OD value.

Allantoic fluid are considered:

- *positive* when producing reactivity ≥ 0.2 OD.
- *negative* when producing reactivity < 0.2 OD

SOP 20. Ribonucleic Acid Extraction from Samples Using TRIzol Reagent

Courtesy of Donald King and Scott Reid

Institute of Animal Health, Pirbright Laboratory, United Kingdom

Responsible Scientists: Donald King (Donald.king@bbsrc.ac.uk) and Scott Reid (Scott.reid@bbsrc.ac.uk)

1 Introduction

1.1 Purpose

This SOP describes a simple and reliable extraction technique using commercial reagents (TRIzol) to provide RNA for subsequent reverse transcription and specific PCR amplification. Samples suitable for testing include field samples and cells suspected to contain RNA viruses of vesicular disease. Extraction of total RNA from such samples is necessary to facilitate the processes of reverse transcription and polymerase chain reaction (RT-PCR) amplification.

This SOP was reviewed in August 2008. This version of the SOP was prepared for IAEA.

2 Risk assessment

If this SOP is not followed, the risks include, but are not limited to:

- Compromising the Quality Management System.
- Failures in the Quality Management System go undetected.
- Failures that are detected, are not corrected or there is undue delay.
- The importance of the processes is not recognised.
- Failing to deliver product, e.g. scientific publication, new diagnostic tests, reports, etc., that is fit for purpose and not in keeping with customer requirements.
- Lack of traceability of working practices.
- Lack of credibility of data.
- Lack of accountability and responsibility.

Note: This procedure does not involve handling of inactivated virus. Work with inactivated viruses (such as the processing of samples and preparation of lysed material suitable for RT-PCR analysis) is described in further IAH documentation.

2.1 Health and Safety: and Other Risks

Under the Management of Health and Safety at Work Regulations 1999 a suitable and sufficient assessment of the risks to the health and safety of employees must be carried out on all work procedures, *e.g.* SOP, method, protocol. Control measures must be put in place to minimise any hazards associated with the work that may endanger the health and safety of employees, as far as reasonably practicable, and consideration should be given to including these control measures in the SOP. Only trained, competent and qualified personnel may perform the procedures as documented in training records.

- Disposable gloves and a laboratory coat should be worn to avoid contact with skin and clothes when using TRIzol Reagent as it contains phenol. Tubes containing waste TRIzol Reagent must be placed into a suitable container for appropriate disposal.
- RNA extraction must be carried out in a location separate from areas where reverse transcription and PCR amplification are carried out to avoid contamination of stock reagents (SEE accompanying SOP outlining procedure for use of Molecular Diagnostic Suite).
- Disposable gloves must be worn throughout the technique as RNases from the skin have an adverse affect on RNA. Precipitated extracted RNA pellets may be very small or invisible; care must be taken to avoid accidental discard. Extracted RNA should be stored at –90 to –50°C in a monitored freezer.

Additional issues that may need to be considered include, but are not limited to:

- Pathogens infective to humans.
- Use of Personal Protection Equipment (PPE).
- Cat III handling.
- Environmental impact.
- Animal welfare.
- Receipt of parcels.
- COSHH.
- Generation of aerosols (by centrifugation, sonication, *etc.*).
- Large volumes (in excess of 100ml) of virus/liquid waste produced

3 Responsibilities

3.1. Personnel using this SOP are responsible for ensuring that they have read and understood the contents, and the procedure is followed. They must be suitably trained and competent, as documented in their training records. It is the duty of the Head of Division/Department/Group (or nominee) to ensure that staff are aware of this responsibility.

3.2. Health and Safety: This SOP should be used in accordance with the current Health and Safety Policy and other relevant instructions as issued by the Institute for Animal Health. Due consideration must be given to National standards and regulations.

4 Materials

Refrigerated microfuge	Heraeus Biofuge *fresco* or one of similar performance
Vortex mixer	Clifton Cyclone or one of similar performance
Vacuum pump	L2C CAPEX (Charles Austen Pumps) or similar
Disposable gloves	medical examination gloves, any supplier
Tubes and caps	2ml skirted and 1.5 ml conical, sterile (Sarstedt or similar)
Calibrated pipettes	single channel, volumes as appropriate, any supplier
Aerosol resistant tips	Molecular Bioproducts ART[R] or similar
Refrigerator	+1 to 8°C (PCR +4°C)
Freezer	–30 to –5°C (PCR –20°C Number 1)
Freezer	–90 to –50°C (top shelf, "Diagnostic PCR samples only" in WRL 50)
Medical sharps container	Cinbin or similar

5 Chemicals

Disinfectant	See current Disease Security Manual, store at approximately 22°C
TRIzol Reagent	Invitrogen, stored between +1 and 8°C (1 ml aliquots in refrigerator PCR +4°C in WRL12a
Chloroform	BDH Analar, store at approximately 22°C
Isopropyl alcohol	BDH Analar, store at approximately 22°C
Ethanol	BDH Analar, store at approximately 22°C

6 Reagents

0.2 M Glycogen	Roche or similar, store between –30 and –5°C, drawer 3 in freezer PCR CLEAN ROOM –20°C in WRL12a
Nuclease-free water	Promega or similar, store at approximately 22°C
Sterile distilled water	Store at approximately 22°C

7 Media

Not applicable

8 Organisms

Not applicable

9 Documentation

This SOP is a component of a larger laboratory system that describes an integrated approach for the diagnosis of vesicular diseases. Therefore, in order to undertake receipt, testing and reporting of samples, it may be necessary to consult further SOP documentation that describes these activities.

10 Procedure

10.1 Ensure any samples frozen and previously stored in TRIzol are thawed. Incubate for 5 min at approximately 22°C (room temperature).

10.2 Label a fresh 2 mL skirted Sarstedt tube corresponding to each test sample to be processed; add 200 μL of chloroform to each 2 mL tube followed by 1.0 mL of the TRIzol solution containing the sample.

10.3 Vortex mix each tube for 10–15 s.

10.4 Centrifuge the tubes for 15 min at 13,000 rpm at 2–8°C.

10.5 Label a fresh 1.5 mL conical tube for each test sample and add 1.0 μL of glycogen to each.

10.6 Remove 500 μL of the top phase (clear phase) from the centrifuged tubes (from **stage 5.4**) and add to the corresponding glycogen-containing tubes. Add 500 μL of isopropyl alcohol to each tube.

10.7 Vortex mix each tube for a few seconds.

10.8 Incubate the tubes at approximately 4°C (on ice) for 10 min.

10.9 Orientate the tubes in the centrifuge using the tube label so that the RNA pellet will be in an expected position. Centrifuge the tubes for 10 min at 13,000 rpm at 2–8°C.

10.10 Discard the supernatant from the tubes and then add 1.0 mL of 70% ethanol to each one.

10.11 Vortex mix the tubes for a few seconds.

10.12 Orientate the tubes in the centrifuge as in **stage 5.9** and centrifuge for 10 min at 13,000 rpm at 2–8°C.

10.13 Remove as much supernatant as possible using gentle vacuum suction (or pipetting) without disturbing the RNA pellet. Air dry each tube for 2–3 min at approximately 22°C. Resolubilise each RNA pellet by adding 20 μL of nuclease-free water to each tube and gentle mixing.

10.14 Maintain the RNA samples on ice and immediately proceed with RT-PCR or store them at –90 to –50°C until required.

11 Storage of Samples

RNA samples should be stored for at least 3 months to allow the one-step RT-PCR assay to be repeated (if necessary). These should be kept at –90 to –50°C on the top shelf of the "Diagnostic PCR samples only" freezer).

12 Results

Results will only become known after analysis of the subsequent reverse transcription and PCR amplification procedures (See SOP: ***Operation of the Stratagene Mx4000/Mx3005P for real-time PCR. One-step RT-PCR amplification of RNA from vesicular disease viruses***).

13 Troubleshooting

Not applicable: Results will only become known after analysis of the subsequent reverse transcription and PCR amplification procedures (see reference to these subsequent SOPs for guidance)

14 Bibliography

1. Chomczynski, P, Sacchi, N. 1987. Single-step method of RNA isolation by acid guanidinium thiocyanate-phenol-chloroform extraction. *Anal. Biochem.* 126, 156–9.
2. Simms, D, Cizdziel, PE, Chomczynski, P. 1993/1994. TRIzol: a new reagent for optimal single-step isolation of RNA. *Focus* 15, 99–102.

15. Tracking Sheet Used for Manual (TRIZOL) Extraction of RNA

Molecular Diagnostics Tracking Sheet

QAU Form 233_03/06

Manual RNA extraction/manual one step RT-PCR for Vesicular viruses

TEST NUMBER: [ ]

SAMPLE DETAILS:

1	9	17
2	10	18
3	11	19
4	12	20
5	13	21
6	14	22
7	15	23
8	16	24

STEP 1: RNA EXTRACTION [WRL 021]

Operator: [ ]

Date: [ ]

Reagent	Batch numbers	vol (ηL)	Pipette serial numbers
TRIZOL reagent	[ ]	1000	[ ]
Chloroform	[ ]	200	[ ]
Glycogen	[ ]	1	[ ]
Aqueous phase		500	[ ]
Isopropanol	[ ]	500	[ ]
Ethanol	[ ]	1000	[ ]
Nuclease-free H_2O	[ ]	20	[ ]

STEP 2: RT-PCR SET-UP [WRL 026]

Operator: [ ]

Number of wells: [0]

Date: [ ]

Reagent	Batch numbers	vol (ηL)	Pipette serial numbers
2x reaction mix	[ ]	12.5	[ ]
SSIII/Taq mix	[ ]	0.5	[ ]
Nuclease-free H2O	[ ]	1.5	[ ]

	vol (ηL)	FORWARD	vol (ηL)	REVERSE	vol (ηL)	PROBE	
☐ FMDV - 5'UTR	2	[ ]	2	[ ]	1.5	[ ]	[ ]
☐ FMDV - 3D	|	[ ]	|	[ ]	|	[ ]	[ ]
☐ SVDV - 2B-IR	|	[ ]	|	[ ]	|	[ ]	[ ]
☐ VESV	|	[ ]	|	[ ]	|	[ ]	[ ]

Plate spin (1 min @ 1,000 rpm) ☐ (Check box)

STEP 3: THERMOCYCLING [EQL 790]

Operator: [ ]

Thermocycler used: [ ]

Date: [ ]

Results file name: N:/Vesicular Dis Grp/quality/Archive/Molecular biology/pcr/RT-PCR results/ [ ]

SOP 21. Ambion Magnetic Beads Extraction (96-well)

Courtesy of Sharon Hietala and Beate Crossley
California Animal Health and Food Safety Laboratory
University of California
Davis
United States of America.

Email: skhietala@ucdavis.edu

1 Purpose

This procedure provides instructions for high-throughput 96-well RNA extraction using micro-sphere bead-based RNA isolation technology manufactured by Ambion/Asuragen Inc.

2 Scope

This procedure follows the work-instructions for loading 96-well plates (DBio_02-043). After completion, a quantitative reverse transcription real time PCR procedure is followed (for example, DBio_02-011 or DBio_02-012).

3 Definitions and Acronyms

qRT PCR, quantitative/real time reverse transcriptase polymerase chain reaction

4 Specimen Information

1. Tissue culture fluid, allantoic fluid, swab fluid, or media (50 uL) pre-loaded in a 96-well plate per DBio_02-043.
2. Loaded plates are stored at 4°C for up to 4 h, or frozen at –20°C until use.
3. Plates that show evidence of spilling or contamination in the form of visible droplets on the lid or on the edge of the plate must be discarded in 10% bleach.

5 Reagents and Media

Media
MagMax-96 (Ambion #1835 or 1836-4 or equivalent)
Kit components are prepared according to manufacturer's instructions (DBio_02-048)
Isopropanol, 100%
Ethanol, 100% analytic grade

6 Supplies and Equipment

Magnet Stand (Ambion, #10027 or 10050)
Vortex GenieII or equivalent (8-speed settings)
Titer Plate Shaker (Lab-Line Instrument Inc or equivalent)
Matrix Impact Equalizer 12-channel 250 μL and 1250 μL pipettors or equivalent
Disposible pipet tips (250 μL, 1250 μL)
96-well polystyrene round bottom plates (Evergreen 222-8032-R1K or equivalent)

7 Special Safety Precautions

Extraction procedure has to be performed in a Class 2 biosafety cabinet until the lysis buffer is added. After addition of the lysis buffer, the extraction process can be completed on the bench using standard biosafety level 2 practices (personal protective wear, autoclave disposal of waste, and chemical decontamination of area).

8 Equipment Calibration and Maintenance

Regular maintenance, see DBIO_02-038 and DBio_02-036.

9 Quality Control

1. RNA extraction efficiency is monitored during PCR using a positive control sample included with each assay run or batch. Performance limits of the positive extraction control are noted in the QC requirements for the individual diagnostic PCR assays.

2. Quality Control values are recorded using the corresponding Quality Control tracking Excel sheets. A QC form accompanying each assay run or batch is used to record assay information unique to the specific run for QC purposes; including reagent lot-numbers, control IDs, performing technician, Quality Control values, and any technical observation made during the assay set-up.
3. Temperatures and volume measurements provided in the SOP are within the calibration tolerance of the respective equipment. Performance times recommended for individual procedure are approximate unless specifically noted.

10 Test Method Instructions

10.1 General Recommendations

- To remove fluid program pipettor to slow speed, add fluid with pipettor programmed on medium speed.
- To remove fluid leave plate on magnetic stand; To add fluid remove plate from the magnetic stand.
- Beads should be suspended after adding elution solution

10.2 Total RNA Isolation

Place bead solution on vortex and shake on setting 3 for two min.

If samples have been pre-plated and the 96-well plate frozen, thaw in a laminar flow hood with the lids off to prevent condensation from running onto plate surface.

Double check the 96-well plate for loading errors (empty wells, excessive volume in wells), and for missing plate numbers/missing bar codes.

10.2.1 RNA Binding

1. Add 101 μL Lysis/Binding solution into each well.
2. Shake for 5 min – shaker dial position 5.
3. Fill 1250 μL/dispense 101 μL
4. Add 20 μL beads/binding mix in each well of a U-bottom plate.
5. Shake for 4 min – shaker dial position 5.
6. Fill 250 μL/dispense 20 μL.
7. Move the U-bottom plate onto the Magnetic Stand, wait 2 min for beads to pellet. Slowly remove supernatant.
8. Fill 240 μL/dispense 240 μL.

Add 100 μL prepared Wash Solution I, shake for 1 min (it is possible that the beads won't break up, there is not a negative effect on RNA quality or yield).

Shaker dial position 5.5.

Fill 1250 μL/dispense 100 μL.

Move the U-bottom plate on to the Magnetic Stand, wait 1 min to pellet beads. Remove supernatant.

Fill 240 μL/dispense 240 μL.

Add 100 μL prepared Wash Solution II, shake for 30s.

Shaker dial position 5.5

Fill 1250 μL/dispense 100 μL.

Move the U-bottom plate on to the Magnetic Stand, wait 30 s to pellet beads. Remove supernatant.

Fill 240 μL/dispense 240 μL.

Add 100 μL prepared Wash Solution II

Shake for 30 s -shaker dial position 5.5.

Fill 1250 μL/dispense 100 μL.

Move the U-bottom plate on to the Magnetic Stand, wait 30 s to pellet beads. Slowly remove supernatant.

Shake plate for 2 min dry to help the evaporation of EtOH (part of the washing solution) – shaker dial position 9.

10.2.2 RNA Elution

1. Add 30 μL Elution Solution to each well, shake for 4 min. Beads should be suspended, otherwise use a manual 200 μL pipettor to resuspend the beads by pipetting up and down. Shaker dial 9.
2. Fill 250 μL/dispense 30 μL.
3. Transfer plate to chemistry room (rm#1213), store in the refrigerator underneath the BL-2 hood until transfer of RNA to RT-PCR master mix.
4. If the RNA will not be used for PCR within 4 h, transfer the RNA into a clean plate (place the U-bottom plate on the Magnetic Stand, wait 1 min to pellet the beads.
5. Transfer the supernatant into a clean plate (fill 25 μL /dispense 25 μL), cover with an adhesive plate sealer, wrap with aluminum foil and freeze the plate at –70°C.)

11 Calculations

N/A: protocol provides sample for subsequent PCR assay.

11.1 Expected Values

N/A: protocol provides sample for subsequent PCR assay.

11.2 Interpretation of Results

N/A: protocol provides sample for subsequent PCR assay.

11.3 Method Limitations

Validated only for samples identified.

11.4 Post Analytical Procedures

RealTime PCR, see DBio_02-011 and DBio_02-012

12 Bibliography and Related Documentation

1. http://www.ambion.com/techlib/posters/thruput_mag_0402.htmL (as of 03_2005).
2. http://www.ambion.com/techlib/prot/fm_1830.pdf (as of 03-2005).
3. Crossley, et al., High-throughput realtime RT PCR assay to detect exotic Newcastle disease virus during the California 2002–2003 outbreak.

13 Appendices

Work instructions (didacted protocol).

14 Purpose

Work instruction for 96-well bead volume, time, and shaker speed (SOP = DBio-02-005).

15 Procedure

1. Transfer 50 μL sample into each well.
2. Add 101 μL *Lysis/Binding Mix* to inactivate virus.
3. Add 20 μL *Beads/Binding Mix* and shake 4 min.
   Shaker dial position 5.
4. Pellet beads 2 min and remove supernatant.
5. Add 100 μL *Wash Solution I Mix* and shake for 1 min.
   Shaker dial position 5.5.
6. Pellet beads 1 min and remove supernatant.
7. Add 100 μL *Wash Solution II Mix* and shake 30 s.
   Shaker dial position 5.5.
8. Pellet beads 30 s and remove supernatant.
9. Add 100 μL *Wash Solution II Mix* and shake for 30 s.
   Shaker dial position 5.5.
10. Pellet beads 30 s and remove supernatant.
11. Shake vigorously for 2 min to briefly dry the beads.
    Shaker dial position 9.
12. Add 30 μL *Elution Solution* (room temp) and shake for 4 min.
    Shaker dial position 9.
13. Pellet beads for 2 min and transfer RNA into storage plate or qRT PCR plate.

SOP 22. Svanodip® FMDV-Ag Penside Test

LAB-ON-SITE
EU PROJECT
SSPE-CT-2004-513645

Courtesy of Ann Nordengrahn
SVANOVA Biotech AB,
Dag Hammarskjölds väg 32A
751 83 Uppsala,
Sweden

Responsible scientist Ann Nordengrahn
E-mail ann.nordengrahn@svanova.com

Created 2008-02-06

1 Introduction

The FMDV-Ag test is a simple direct test for the detection of all seven serotypes of the FMDV antigen in clinical samples and can be carried out at the location of the farm. In this assay FMDV antigen specific antibodies have been bound to colloid gold as well as immobilised on the membrane. If present in the sample, FMDV antigen binds to the gold conjugate forming an immune complex. The complex then migrates by capillary action along the membrane until it reaches the immobilised antibody. The complex will bind to the immobilised antibody, resulting in an accumulation of colloid gold (a red/purple line) visible by eye. A band in the Test (T) window indicates a positive result. No band in the Test (T) window indicates a negative result. The antibody in the Control (C) window always binds colloid gold-antibody, regardless of FMDV viral presence. This C-band ensures correct test performance.

2 Contents

The Svanodip® FMDV-Ag penside test can be purchased from Svanova Biotech AB, Uppsala Science park, Sweden. +4618654900.

2.1 The Kit Contains

- 20 Test devices pre-packed in sealed foil bags with desiccant.
- Bottles with dropper containing sample dilution buffer.
- Plastic pipettes.
- Test (Eppendorf) tubes with lid.
- Swabs (two in each pack).

2.2 The Sample Preparation Kit Contains

- 20 Bottles with ball bearings.
- Bottle containing sample dilution buffer.
- Plastic pipettes.

3 Procedure

3.1 Specimen Collection and Processing

3.1.1 Vesicular Fluid

A small volume of vesicular fluid is collected by a syringe and the amount is calculated.
An equal amount of sample dilution buffer is added to the sample and mixed.

3.1.2 Epithelium

Approximately 0.2 g of epithelium (the size of the nail on a little finger) is collected from the surface or margins of vesicles.

Best results will be achieved with fresh and friable material.

Add 2 mL of sample buffer into the glass bottle.

Place the epithelium into the glass bottle and vigorously shake four times over a 5 min period, with the aim of liquidizing the material.

3.2 Precautions

1. Carefully read and follow all instructions.
2. Store the kit and all reagents at +4°C to +28°C (39 to 82°F).

3. All reagents should equilibrate to room temperature 18°C to 25°C (64 to 77°F) before use.
4. Handle all materials as though capable of transmitting infectious disease.
5. Do not mix components or instruction booklets from different test kits.
6. Care should be taken to prevent contamination of kit components.
7. Do not use test kit beyond expiratiopn date.
8. Do not eat, drink, or smoke where specimens or kit reagents are handled.
9. Use a separate pipette tip for each sample.
10. Burn all unused biological materials before disposal.

3.3 Procedure

1. Make sure that the device is at ambient temperature before opening the bag.
2. Remove the test device from the bag and place on a flat surface.
3. Use the pipette to take prepared sample solution (see point 3.1) up to the mark on the pipette. Hold vertically and slowly add five drops (approximately 200 μL) of the sample solution to position "S" on the test device.
4. Read results within 5–10 min. Disregard any lines which appear after 45 min.

4 Interpretation of the Results

1. A line in the Control (C) window shows that the test has worked correctly. Any device that has not produced a control line at 10 min must be classified as void.
2. A line in the Test (T) window indicates a positive result and that the sample contains FMDV antigen. If no line appears in the Test (T) window the test is negative for FMDV antigen (see figure below).
3. A difference in the intensity may occur between the line in the Test (T) and Control (C) window but this does not affect the interpretation of the result.

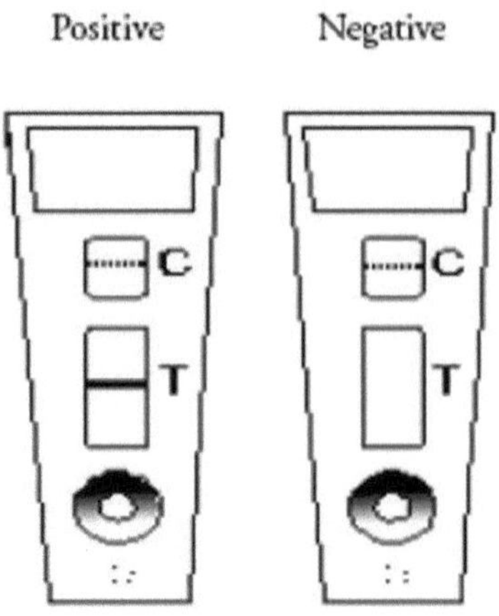

SOP 23. FMDV PLA Assay

LAB-ON-SITE
EU PROJECT
SSPE-CT-2004-513645

Courtesy of Ann Nordengrahn
SVANOVA Biotech AB,
Dag Hammarskjölds väg 32A
751 83 Uppsala,
Sweden

Responsible scientist Ann Nordengrahn
E-mail ann.nordengrahn@svanova.com

Created 2008-02-06

1 Introduction

The Proximity ligation assay (PLA) can be used for detection of different pathogens such as FMDV virus. The basis of the PLA is that pathogen specific-antibodies binding target proteins are coupled to non-sense oligonucleotide strands (proximity probes). These proximity probes can be joined by ligation when two or more such reagents are brought into proximity by binding to the same target molecule or target molecule complex (Fig. 5.2). The DNA ligation products are subsequently detected by PCR amplification using fluorogenic probes to detect the amplified product. The assay shows a high sensitivity and specificity that is in line with RT PCR. For more information about the assay development and evaluation on clinical samples see Nordengrahn et al., Vet Microbiol 2008, 18;127 (3–4), 227–236.

2 Preparation of Proximity Probes

2.1 Biotinylation of Antibody

1. D-biotin-N-hydroxysuccinimide ester (Roche Diagnostics Corp. Germany) is mixed with the antibody in a 10-fold molar excess and with a volume ratio of 1:10.
2. The solution is left for 4 h incubation at room temperature and at constant agitation.

3. The biotinylated antibody is dialysed against Phosphate buffered saline (PBS, pH 7.4) and stored at –20°C until use.

2.2 Construction of Proximity Probes

1. Streptavidin conjugated oligonucleotides 3′ and 5′ (Olink Bioscience, Uppsala Sweden) are mixed with the biotinylated antibody; 30 nmol/L antibody is mixed with 30 nmol/L oligonucleotide 3′ and 5′ respectively in a final volume of 5 μL and left for 1 h at room temperature.
2. The proximity probes are diluted to a final concentration of 1.2 nmol/L in a probe dilution buffer and stored at +4°C until use.

2.2.1 Probe Dilution Buffer

PBS, 10 g/L Bovine serum albumin (BSA), 16 mg/L sheared polyA bulk nucleic acid (Sigma), 1 mmol/L D-biotin (Molecular Probes)

3 Assay Procedure

3.1 Detection of FMDV by Homogenous Phase PLA

1. Dilute the samples 1 in 5 in PBS (or any other dilution factor of choice).
2. Mix 1 μL of diluted sample with 4 μL of a solution containing both proximity proximity probes (each diluted to a concentration of 24 pM in probe dilution buffer) in optical PCR tubes (Applied Biosystems, US) and incubate for 1 h at 37°C.
3. Add 50 μL of the ligation and TaqMan PCR mix and incubate for 5 min at room temperature.

Ligation and TaqMan mix

50 mM KCl, 10 mM Tris-HCl pH 8.3, 3.15 mM $MgCl_2$, 0.4 Weiss units of T4 DNA ligase (Fermentas), 400 nM connector oligonucleotide (Olink Bioscience, Uppsala Sweden), 80 μM ATP, 200 μM each of the deoxynucleoside triphosphates, 100 nM primers forward and reverse (Olink Bioscience, Uppsala Sweden), 100 nM TaqMan® MBG probe (Applied Biosystems), 1.5 U Platinum Taq DNA polymerase (Invitrogen).

4. Transfer the tubes to a real-time PCR instrument and use the following temperature profile, 95°C 2 min, 45 cycles, 95°C 15 s and 60°C 2 min.

4 Interpretation of Results

The assay cut-off value is set to two standard deviations (*S.D.*) over the background signal (CT value).

Samples with values below this threshold are considered to be negative while samples with values higher than the threshold are positive.

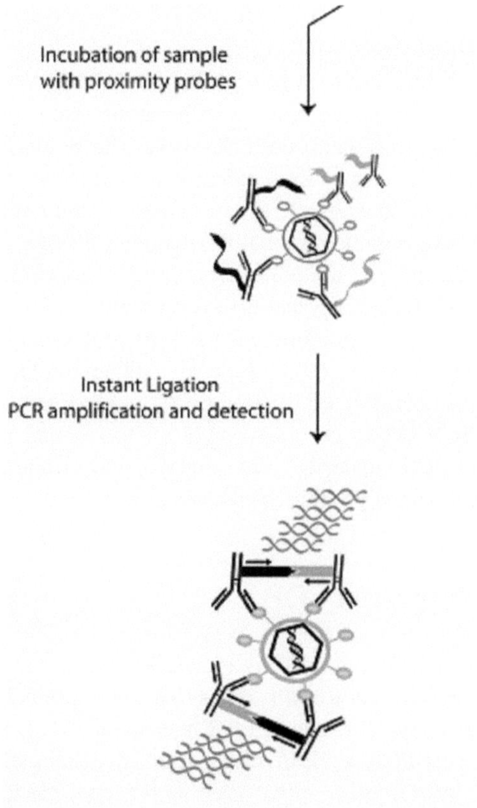

Fig. 5.2 One step proximity ligation and detection

This figure has been published by Nordengrahn et al., in *Vet. Microbiol.* 2008, 18;**127** (3–4), 227–36.

SOP 24. Procedure for Using the Molecular Diagnostics Suite

Courtesy of Donald King and Scott Reid

Institute of Animal Health, Pirbright Laboratory, United Kingdom

Responsible Scientists: Donald King (Donald.king@bbsrc.ac.uk) and Scott Reid (Scott.reid@bbsrc.ac.uk)

1 Introduction

Molecular assays such as polymerase chain reaction (PCR) are important tools for diagnostic and research work. These assays are highly sensitive and are able to detect only a few copies of the target template in a sample. Unfortunately, methods such as PCR are prone to contamination, which can generate false positive results in the downstream assays. The most common source of contaminating template is previously amplified material which is transferred by aerosols, clothing, hands or equipment.

In order to minimize the risk of contamination, it is important that the various steps of the PCR process (TEMPLATE RNA/DNA EXTRACTION – MASTERMIX PREPARATION – ASSEMBLY OF PCR REACTION – POST PCR WORK) are segregated. In particular, care should be taken not to bring post-PCR material into contact with un-amplified template or any of the components of the PCR master-mix.

2 Purpose of SOP

The aim of this SOP is to outline the correct use of the molecular diagnostics suite. This suite comprises 4 dedicated rooms for (1) template RNA/DNA extraction, (2) preparation of PCR master-mixes, (3) assembly of PCR reactions and (4) all post-PCR work involving amplified samples and other high-risk material containing target template such as bacterial clones and plasmids. The location of these rooms is shown in Fig. 5.3. Together, the aim of this suite is to *minimise* the risk of the occurrence of PCR contamination.

This SOP was reviewed in August 2008. This version of the SOP was prepared for IAEA.

3 Risk Assessment

If this SOP is not followed, the risks include, but are not limited to:

- Compromising the Quality Management System.
- Failures in the Quality Management System go undetected.

- Failures that are detected, are not corrected or there is undue delay.
- The importance of the processes is not recognised.
- Failing to deliver product, e.g. scientific publication, new diagnostic tests, reports, etc., that is fit for purpose and not in keeping with customer requirements.
- Lack of traceability of working practices.
- Lack of credibility of data.
- Lack of accountability and responsibility.

Note: no inactivated virus is allowed in the template extraction room, master-mix preparation room and PCR assembly room. Use of inactivated viruses is permitted in the post-PCR room under the guidelines outlined in BSEC-REG-019

3.1 Health and Safety: and Other Risks

Under the Management of Health and Safety at Work Regulations 1999 a suitable and sufficient assessment of the risks to the health and safety of employees must be carried out on all work procedures, e.g. SOP, method, protocol. Control measures must be put in place to minimise any hazards associated with the work that may endanger the health and safety of employees, as far as reasonably practicable, and consideration should be given to including these control measures in the SOP. Only trained, competent and qualified personnel may perform the procedures as documented in training records.

Additional issues that may need to be considered include, but are not limited to:

- Pathogens infective to humans.
- Use of Personal Protection Equipment (PPE).
- Cat III handling.
- Environmental impact.
- Animal welfare.
- Receipt of parcels.
- COSHH.
- Generation of aerosols (by centrifugation, sonication, etc.).
- Large volumes (in excess of 100ml) of virus/liquid waste produced

3.2 Responsibilities

3.3. Personnel using this SOP are responsible for ensuring that they have read and understood the contents, and the procedure is followed. They must be suitably trained and competent, as documented in their training records. It is the duty of the Head of Division/Department/Group (or nominee) to ensure that staff are aware of this responsibility.

3.4. Health and Safety: This SOP should be used in accordance with the current Health and Safety Policy and other relevant instructions as issued by the Institute for Animal Health. Due consideration must be given to National standards and regulations.

4 Materials

4.1 Personal protective equipment to be worn while working in the rooms and reagents required.

4.1.1 *Template Extraction Room*: Dedicated Laboratory Coat (WHITE) and gloves (any disposable).

4.1.2 *Master-Mix preparation Room*: Dedicated Laboratory Coat (YELLOW), Gloves (any disposable), Overshoes (Scientific Laboratory Supplies).

4.1.3 *PCR Assembly Room*: Gloves (any disposable), DNAzap1 and DNAzap2 (Ambion) or similar reagent.

4.1.4 *Post-PCR Room*: Dedicated Laboratory Coat (BLUE), Gloves (any disposable) and Overshoes (Scientific Laboratory Supplies).

5 Procedure

5.1 Template Extraction Room

Function of room: This room is used to prepare RNA/DNA template for the diagnostic assays to detect foot-and-mouth disease virus (FMDV), swine vesicular disease virus (SVDV), African swine fever virus (ASFV), bluetongue virus (BTV) and Rinderpest. Other areas may be suitable for template preparation for other uses. If other areas are used, their use should, where possible, follow guidelines similar to those outlined in this document. (NB: the MagNA PURE LC robot located in this room is also used to perform PCR set-up steps with cDNA template).

5.1.1 Use of Room

STEP 1: Before entering room, discard any gloves that you are already wearing

STEP 2: Put on dedicated Laboratory coat (this coat should NOT be worn outside of this room).

STEP 3: Put on fresh gloves (if required) before using the equipment in the room. Change gloves frequently, especially if they are suspected of becoming contaminated with solutions containing template DNA/RNA.

STEP 4: Work areas should be cleaned up after use. In the event of a spillage involving template RNA/DNA, the workstation should be thoroughly cleaned using DNAzap (or similar suitable product).

5.2 Master-Mix Preparation Room

Function of Room: This room should be used for the storage of all the components (except RNA/DNA template) of reverse transcriptase and PCR master-mixes. The room contains OMNI PCR workstations that are used for the preparation of reaction master-mixes. These units contain main and exhaust HEPA filters providing a clean airflow over the work surface and providing protection to the operator. These units meet the clean air requirement to ISO 14644-1 1999 class 6.

5.2.1 *Entry/Exit* into Room.

Entry:

STEP 1: Remove any lab coat and discard gloves already being worn. Hooks for labcoats are available outside of the *PCR assembly room.*

STEP 2: Enter Lobby

STEP 3: Put on Gloves

STEP 4: Put on Dedicated YELLOW labcoat

STEP 5: Open inner door, Put on overshoes as the threshold boundary is crossed

Note: Do not bring anything into room, including protocols and lab-books, ice buckets, racks, tubes and RNA/DNA template.

Exit:

STEP 1: Open Inner door

STEP 2: If required, place tubes containing reaction mix on the other side of threshold boundary.

STEP 3: Step out of room, removing overshoes and placing them into recycle bin. These overshoes can be recycled for use in the Post-PCR room (see below).

STEP 4: Remove dedicated labcoat and glove(s).

STEP 5: Exit Lobby with the prepared master-mix

5.2.2 *PREPARATION* of Reaction Master-Mixes

These can be prepared according to your SOPs or methods sheets. These can be viewed and printed using the Networked computer terminal in the room. Master-mixes should be prepared in the class I hoods using the dedicated micro-pipettes (with aerosol-resistant pipette tips) for all liquid handling steps. Chilled master-mixes can be prepared using 96-well aluminium blocks which are stored in the freezer (these blocks should be returned to freezer after use).

Note: Gloves should be worn at all times while in the room

5.2.3 *RECEIVING* New Materials for Storage in Freezers/Fridges in Master-Mix Preparation Room

STEP 1: One person to enter room (as above)
STEP 2: Second person to remain in the lobby and open all secondary/external packaging
STEP 3: Person inside room should reach over the threshold boundary and take hold of the items to be brought into the room. Care should be taken not to touch the outside packaging.
STEP 4: Material can then be stored in fridge or freezer (as appropriate).

Note: Where possible, no external packaging should be brought into the room

5.3 PCR Assembly Room

Function of Room: This room should be used for the final assembly of reverse transcription or PCR reactions. RNA/DNA template is added to the reaction master-mixes (prepared as above). This activity can be performed manually using the dedicated micro-pipettes or using suitable protocols on the Qiagen BR3000. The intended use for this room is restricted to un-amplified RNA/DNA template (i.e. that obtained directly from clinical samples and similar material). If RT or PCR is to be performed on previously amplified template (such as that derived from PCR products or plasmids), then these reactions should be assembled in the Post-PCR room (see below).

NB: The pipettes supplied in the room should NOT be used for work involving RNA standards (such as those used in quantitative RT-PCR applications). A separate room with different micro-pipettes should be used for the manual transfer of this type of material.

5.3.1 Use of Room

STEP 1: Discard any gloves already being worn.
STEP 2: Put on fresh gloves (if required) before using the equipment in the room. Change gloves frequently, especially if they are suspected of becoming contaminated with solutions containing template DNA/RNA.
STEP 3: Area should be cleaned up after use. In the event of a spillage involving template RNA/DNA, the workstation should be thoroughly cleaned using DNAZap (or similar suitable product). DNAZap consists of two solutions that are innocuous by themselves, but which become a potent nucleic acid degrading solution when mixed. This mixture is able to degrade high levels of contaminating DNA and RNA from surfaces instantaneously. All contaminating nucleic acid is degraded to nucleotides, preventing any chance of false positive amplification. The use of DNAZap is outlined in the protocol from Ambion (http://www.ambion.com/techlib/prot/bp_9890.pdf). Briefly, spray or apply *Solution 1* on the surface that requires cleaning followed by *Solution 2*. Wipe with clean paper towel, rinse twice with distilled water and wipe dry with clean paper towel.
NB: Wear Gloves at all times since Solutions 1 and 2 are acidic/corrosive, can cause irritation and are harmful if swallowed.

5.4 Post-PCR Room

Function of Room: This room is to be used for any laboratory activity involving post-PCR material. Examples of this type of work include gel electrophoresis, clean-up of PCR and restriction enzyme fragments from gels, cloning and preparation of sequencing reactions. These activities are segregated from other laboratory work in an attempt to minimise the potential introduction of this post-PCR material into the PCR reactions or contamination of the master-mix components.

5.4.1 Entry/Exit

Entry:

STEP 1: Enter Lobby
STEP 2: Remove any lab coat and discard gloves already being worn.
STEP 2: Put on Gloves (to be worn at all time)
STEP 3: Put on Dedicated BLUE labcoat
STEP 4: Open inner door, Put on overshoes as the threshold boundary is crossed

Exit:

STEP 1: Open Inner door
STEP 2: Step out of room, removing overshoes and placing them into bin.
STEP 3: Remove dedicated labcoat.
STEP 4: Remove and discard gloves
STEP 5: Exit Lobby

Note: Where at all possible – nothing is to be removed from the post-pcr room

5.4.2 Removal of Waste from Post-PCR Room

All waste is to be placed into tins and autoclaved.

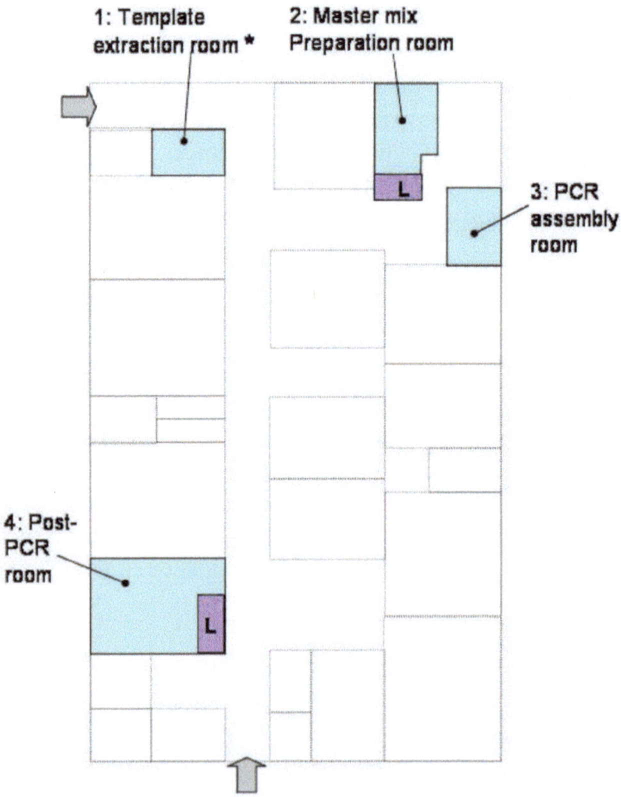

Fig. 5.3 Location of the rooms comprising the Molecular Diagnostics suite in the Epidemiology building. **L** denotes lobby changing areas for master-mix preparation and post-PCR rooms
NB: * other areas maybe appropriate for the preparation of RNA/DNA template by different users

6 Results

No system can completely *eliminate* the risk of cross-contamination during PCR. Therefore, it is important that negative ("no template RNA/DNA") controls which contain all reaction components except the template are included and run in parallel to the samples in the PCRs. Presence of bands (or high fluorescence signal, for real time PCRs) in these samples could indicate that cross-contamination has occurred. In the event of concern regarding cross-contamination during PCR, corrective action listed in the "troubleshooting" section should be implemented

7 Maintenance

7.1 Routine Cleaning

Institute cleaning staff are not allowed to clean the following areas: 1: Master-mix preparation room or 2: Post-PCR room. For these rooms, dedicated cleaning equipment (Floor: mop and detergent, Bench-tops: Cloth and DNAzap) are situated in the rooms. Weekly mopping of the floors is performed by laboratory staff according to a rota. Users of the room are responsible for ensuring that bench-tops are kept clean after use.

7.2 Maintenance of Equipment

Engineers must only enter these rooms as described above in the SOP. Where possible, they should only take essential tools and equipment into these rooms. In common with the scheme for sample management, tools should not be taken directly from the Post-PCR room to other areas in the Epidemiology building.

8 Troubleshooting

The following steps are a guide if cross-contamination of PCR is suspected. This is most usually evident if bands (or fluorescence signal) appears in negative ("no template") control samples.

STEP 1: Repeat PCR assay, in case tubes were mislabelled.
STEP 2: Repeat assay without any template.

STEP 3: Thoroughly clean pipettes in the PCR assembly room and discard all working stocks of primers, dNTPs etc used to make up the master-mix – Repeat Assay.

STEP 4: If a serious degree of contamination is suspected (such as presence of DNA/RNA template in stock PCR master-mix components in the Master-mix preparation room), it is important that other users of the molecular diagnostics suite are informed. Effective corrective action may require the co-ordination between all users of the facility.

9 Bibliography

Not applicable

SOP 25. One step TaqMan® RT-PCR for Diagnosis of FMDV and Related Vesicular Viruses

Courtesy of Donald King and Scott Reid

Institute of Animal Health, Pirbright Laboratory, United Kingdom

Responsible Scientists: Donald King (Donald.king@bbsrc.ac.uk) and Scott Reid (Scott.reid@bbsrc.ac.uk)

1 Introduction

1.1 *Purpose of activity*: To describe the preparation of one-step TaqMan® RT-PCR assays for the amplification of vesicular virus RNA (foot-and-mouth disease virus, swine vesicular disease virus and vesicular exanthema of swine virus).

1.2 *Background*: TaqMan RT-PCR is a sensitive method used for the laboratory detection of vesicular disease viruses. The basis of this method is the use of oligonucleotide primers to specifically amplify target regions of the RNA present on the viral genomes of interest. The assay uses a dual-labelled fluorogenic (TaqMan®) probe which allows the assay to be performed in a closed-tube format minimising the potential for cross-contamination of post-PCR products.

1.3 *Rationale*: PCR amplification of samples is carried out in optical reaction plates. An additional SOP (***Operation of the Stratagene Mx4000/Mx3005P for real-time PCR. One-step RT-PCR amplification of RNA from vesicular disease viruses***) outlines the use of the real-time PCR machines suitable for performing the RT-PCR.

This SOP was reviewed in August 2008. This version of the SOP was prepared for IAEA.

2 Risk Assessment

If this SOP is not followed, the risks include, but are not limited to:

- Compromising the Quality Management System.
- Failures in the Quality Management System go undetected.
- Failures that are detected, are not corrected or there is undue delay.
- The importance of the processes is not recognised.
- Failing to deliver product, e.g. scientific publication, new diagnostic tests, reports, etc., that is fit for purpose and not in keeping with customer requirements.

- Lack of traceability of working practices.
- Lack of credibility of data.
- Lack of accountability and responsibility.

Note: This procedure does not involve handling of inactivated virus. Work with inactivated viruses (such as the processing of samples and preparation of lysed material suitable for RT-PCR analysis) is described in further IAH documentation.

- PCR set up and master-mix preparation must be carried out in dedicated areas to avoid contamination of stock reagents.
- Disposable gloves must be worn throughout the technique as RNases from the skin have an adverse affect on RNA. Precipitated extracted RNA pellets may be very small or invisible; care must be taken to avoid accidental discard. Extracted RNA should be stored at –90 to –50°C in a monitored freezer.

3 Responsibilities

3.1 Personnel using this SOP are responsible for ensuring that they have read and understood the contents, and the procedure is followed. They must be suitably trained and competent, as documented in their training records. It is the duty of the Head of Department/Group (or nominee) to ensure that staff are aware of this responsibility.

3.2 Health and Safety: This SOP should be used in accordance with the current Health and Safety Policy and other relevant instructions as issued by the Institute for Animal Health. Due consideration must be given to National standards and regulations.

3.3 Gloves must be worn whenever PCR plates are handled and must be changed frequently. A clean laboratory coat (dedicated for PCR-related work only) must be worn at all times in the isolated area. Appropriate reagent tracking sheets must be filled in (see examples at the end of this document).

4 Materials

4.1 Equipment (unless otherwise stated, all equipment located in WRL 44/WRL12a)

Disposable gloves	medical examination gloves, any suitable supplier
PCR workstation	ASTEC
Optical reaction plates and caps	Stratagene (Mx4000 or Mx3005P)
PCR plate	ABgene or other suitable supplier
Roller (for secure fastening of optical caps onto plate)	Applied Biosystems
Microcentrifuge tubes	1.5 ml, any manufacturer
Calibrated pipettes	single channel, volumes as appropriate, any manufacturer, dedicated for PCR only
Aerosol resistant tips	Molecular Bioproducts ART[R] or similar
Freezer	–30 to –5°C (PCR –20°C Number 1)
Refrigerator	+1 to 8°C (PCR +4°C)
Large capacity centrifuge	**SORVALL® RC 3C PLUS or one of similar performance, located in WRL 51**
Ice bucket with ice	

4.2 Chemicals

Not applicable

4.3 Reagents

Nuclease-free water	Promega or other suitable supplier, store at +1 to 8°C – PCR clean room
Superscript III RT/Platinum *Taq* 2X reaction mix	Invitrogen, store at –30 to –5°C – PCR clean room
Superscript III RT/Platinum *Taq* Enzyme	Invitrogen store at –30 to –5°C – PCR clean room
Forward primers, reverse primers and fluorogenic probes	Applied Biosystems, Sigma Genosys other suitable supplier, made to specified sequences.

Forward and reverse primers and probes must be diluted to a "working dilution" (see Table 5.3) in nuclease-free water. Stock primers and probe (and additional aliquots of "working dilutions" of primers) are stored at –30 to –5°C, working dilutions are kept at +1 to 8°C.

Table 5.3 Primers and Probes used for One step RT-PCR of vesicular disease viruses

		Oligo name	Sequence (5′–3′)	Working conc (pmols/uL)
FMDV 5′UTR	Forward Primer	SA-IR-219-246F	CAC YTY AAG RTG ACA YTG RTA CTG GTA C	10
	Reverse Primer	SA-IR-315-293R	CAG ATY CCR AGT GWC ICI TGT TA	10
	Taqman PROBE	SAmulti2-P-IR-292-269R	CCT CGG GGT ACC TGA AGG GCA TCC	5
FMDV 3D	Forward Primer	Callahan 3DF	ACT GGG TTT TAC AAA CCT GTG A	10
	Reverse Primer	Callahan 3DR	GCG AGT CCT GCC ACG GA	10
	Taqman PROBE	Callahan 3DP	TCC TTT GCA CGC CGT GGG AC	5
SVDV 2B-IR	Forward Primer	SASVD-2B-IR-252-275F	CGA GAA ACC TAG TAC CAC CAT GAA	10
	Reverse Primer	SASVD-2B-IR-332-312R	CGG TGA CTC ATC GAC CTG ATC	10
	Taqman PROBE	SASVD-2B-IR-289-309P	TCG CTC CGC ACA ACC CCA GTG	5
VESV	Forward Primer	Vesivirus forward primer 1	GAY GAC GGT GTY TAC ATY GTY C	10
	Reverse Primer	Vesivirus reverse primer 1	GGG AYI GGC GTT ATY TCA GCR T	10
	Taqman PROBE	Vesivirus probe 1	CTG AAR CCG ACY CGG ACC GAC A	5

4.4 Media

Not applicable

4.5 Organisms

Not applicable

4.6 Documentation

This SOP is a component of a larger laboratory system that describes an integrated approach for the diagnosis of vesicular diseases. Therefore, in order to undertake receipt, testing and reporting of samples, it may be necessary to consult further SOP documentation that describes these activities.

5 Procedure

5.1 Manual Preparation of One-Step RT-PCR Reaction Plates

5.1.1 Prepare a plan showing the layout of test and control RNA samples on the optical reaction plate. Diagnostic RNA samples should be tested in duplicate.

5.1.2 In the PCR clean room, prepare the one-step RT-PCR reaction mixture described in Table 5.4 in a clean tube. Prepare the reaction mixture in sufficient volume for the total number of samples to be assayed (x) plus one (x+1).

Table 5.4 Composition of reaction mixes for One-step RT-PCR

	Volume added to reaction mix (μL)
Reaction component	For each reaction
2x reaction mix	12.5
Nuclease-free water	1.5
Forward primer	2
Reverse primer	2
TaqMan probe	1.5
Superscript III RT/Platinum Taq Mix	0.5
Total Volume	20

5.1.3 Add 20 μL of the one-step RT-PCR reaction mix to each well of an optical reaction plate for each sample to be assayed. Transfer plate to room WRL10, then add 5 μL of RNA, changing the tip between each sample.

5.1.4 Securely cover each well with a cap in the safety cabinet using a roller.

5.1.5 Spin the covered optical reaction plate for 1 min at 1,000 rpm in the large capacity centrifuge to mix the contents of each well.

6 Results

The results from each step must be printed and appended to the other paperwork for the RT-PCR assay. These documents are archived together in the WRL diagnostic assays folder.

This SOP is part forms part of a complete assay comprising a series of SOPs describing RNA extraction, RT-PCR amplification and analysis and interpretation of the data. The assay performance of the test and control samples can only be judged upon completion of the RT-PCR process (see SOP: ***Operation of the Stratagene Mx4000/Mx3005P for real-time PCR. One-step RT-PCR amplification of RNA from vesicular disease viruses***).

7 Bibliography

1. Oleksiewicz, MB, Donaldson, AI, Alexandersen, S. 2001. Development of a novel real-time RT-PCR assay for quantitation of foot-and-mouth disease virus in diverse porcine tissues. *J. Virol. Methods*, **92**, 23–35.
2. Reid, SM, Ferris, NP, Hutchings, GH, Zhang, Z, Belsham, GJ, Alexandersen, S. 2002. Detection of all seven serotypes of foot-and-mouth disease virus by real-time, fluorogenic RT-polymerase chain reaction assay. *J. Virol. Methods*, **105**, 67–80.
3. Reid, SM, Grierson, SS, Ferris, NP, Hutchings, GH, Alexandersen, S. 2003. Evaluation of automated RT-PCR to accelerate the laboratory diagnosis of foot-and-mouth disease virus. *J. Virol. Methods*, **107**, 129–39.

8 Tracking sheet used for manual (TRIZOL) extraction of RNA

Molecular Diagnostics Tracking Sheet

QAU Form 233_03/06

Manual RNA extraction/manual one step RT-PCR for Vesicular viruses

TEST NUMBER:

SAMPLE DETAILS:

1	9	17
2	10	18
3	11	19
4	12	20
5	13	21
6	14	22
7	15	23
8	16	24

Reagent | Batch numbers | Pipette serial numbers

STEP 1: RNA EXTRACTION [WRL 021]

Operator:

Date:

Reagent	Batch numbers		vol (ηL)	Pipette serial numbers
TRIZOL reagent			1000	
Chloroform			200	
Glycogen			1	
		Aqueous phase	500	
Isopropanol			500	
Ethanol			1000	
Nuclease-free H_2O			20	

STEP 2: RT-PCR SET-UP [WRL 026]

Operator:

Date:

Number of wells: 0

Reagent	Batch numbers	vol (ηL)	Pipette serial numbers
2x reaction mix		12.5	
SSIII/Taq mix		0.5	
Nuclease-free H2O		1.5	

	vol (ηL)	FORWARD	vol (ηL)	REVERSE	vol (ηL)	PROBE	
☐ FMDV - 5'UTR	2		2		1.5		
☐ FMDV - 3D	|		|		|		
☐ SVDV - 2B-IR	|		|		|		
☐ VESV	|		|		|		

Plate spin (1 min @ 1,000 rpm) ☐ (Check box)

STEP 3: THERMOCYCLING [EQL 790]

Operator:

Thermocycler used:

Date:

Results file name: N:/Vesicular Dis Grp/quality/Archive/Molecular biology/pcr/RT-PCR results/

SOP 26. Operation of the Stratagene Mx4000/Mx3005P for Real-Time PCR. One-Step RT-PCR Amplification of RNA from Vesicular Disease Viruses

Courtesy of Donald King and Scott Reid

Institute of Animal Health, Pirbright Laboratory, United Kingdom

Responsible Scientists: Donald King (Donald.king@bbsrc.ac.uk) and Scott Reid (Scott.reid@bbsrc.ac.uk)

1 Introduction

1.1 *Purpose of activity*: To describe the procedure to operate a Stratagene Mx4000 or MX3005P thermal cycler for real-time RT-PCR amplification of RNA for the diagnosis of vesicular and related viruses.

1.2 *Background*: Real-time PCR (TaqMan®) allows the rapid and sensitive detection of viral nucleic acid using PCR, detecting the amplified product by the fluorescence produced due to the cleavage of a sequence-specific probe.

1.3 *Rationale*: The Stratagene Mx4000 and MX3005P instruments carry out the one-step RT-PCR amplification and real-time fluorescence detection of up to 96 samples in optical reaction plates. The basic steps required to set up amplification reactions in the instrument are described in this SOP.

This SOP was reviewed in August 2008. This version of the SOP was prepared for IAEA.

2 Risk Assessment

If this SOP is not followed, the risks include, but are not limited to:

- Compromising the Quality Management System.
- Failures in the Quality Management System go undetected.
- Failures that are detected, are not corrected or there is undue delay.
- The importance of the processes is not recognised.
- Failing to deliver product, e.g. scientific publication, new diagnostic tests, reports, etc., that is fit for purpose and not in keeping with customer requirements.
- Lack of traceability of working practices.
- Lack of credibility of data.
- Lack of accountability and responsibility.

Note: This procedure does not involve handling of inactivated virus. Work with inactivated viruses (such as the processing of samples and preparation of lysed material suitable for RT-PCR analysis) is described in further IAH documentation.

- PCR set up and master-mix preparation must be carried out in dedicated areas to avoid contamination of stock reagents.
- Disposable gloves must be worn throughout the technique as RNases from the skin have an adverse affect on RNA. Precipitated extracted RNA pellets may be very small or invisible; care must be taken to avoid accidental discard. Extracted RNA should be stored at –90 to –50°C in a monitored freezer.

3 Responsibilities

3.1 Personnel using this SOP are responsible for ensuring that they have read and understood the contents, and the procedure is followed. They must be suitably trained and competent, as documented in their training records. It is the duty of the Head of Department/Group (or nominee) to ensure that staff are aware of this responsibility.

3.2 Health and Safety: This SOP should be used in accordance with the current Health and Safety Policy and other relevant instructions as issued by the Institute for Animal Health. Due consideration must be given to National standards and regulations.

3.3 Gloves must be worn whenever PCR plates are handled and must be changed frequently. Appropriate reagent tracking sheets must be filled in (see Appendix 1).

4 Materials

4.1 Equipment (Unless Otherwise Stated, All Equipment Located in WRL 51)

Real-time PCR machine	Mx4000, Stratagene,
	or
Real-time PCR machine	Mx3005P, Stratagene,
Large capacity centrifuge	SORVALL® RC 3C PLUS or one of similar performance
Disposable gloves	medical examination gloves, any suitable supplier

4.2 Chemicals

Not applicable

4.3 Reagents

Not applicable

4.4 Media

Not applicable

4.5 Organisms

Not applicable

4.6 Documentation

This SOP is a component of a larger laboratory system that describes an integrated approach for the diagnosis of vesicular diseases. Therefore, in order to undertake receipt, testing and reporting of samples, it may be necessary to consult further SOP documentation that describes these activities.

Additional documentation that requires consultation:
Stratagene Mx4000 Sequence Detection System User's Manual (2003)
Stratagene Mx3005P Setup and user's guide
MxPro QPCR Software instruction manual

5 Procedure (Alternative Procedures Are Described for the Mx4000 and Mx3005P Machines)

The following procedure is an outline of the steps required to operate the Mx4000 thermal cycler for one-step RT-PCR of FMDV (and other vesicular disease viruses).

5.1 (Mx4000) Switch on the Mx4000 at the back of the instrument. It is advisable to do this some time in advance as it takes time for the computer to link-up with the instrument.

5.2 (Mx4000) Switch on the computer and log on.

5.3 (Mx4000) Open the Mx4000 software by double-clicking on the desktop icon.

5.4 (Mx4000) In "New Experiment Options", select "Quantitative PCR (Multiple standards)" and click OK.

5.5 (Mx4000) If the lamp is switched off (red bulb), switch on the lamp to warm-up by clicking on the bulb on the task bar.

5.6 (Mx4000) Click on the "open door" button on the task bar. Wait for the door to open.

5.7 (Mx4000) Place the PCR plate containing the reactions securely into the 96-well block, ensuring that well A1 is located at the far-left corner.

5.8 (Mx4000) Click on the "close door" button on the task bar. The door will close.

5.9 (Mx4000) In the File menu open the file "WRLprofile Mx4000.mxp" from the N:\Ves Dis Grp\Quality\Archive\Molecular Biology\PCR\WRLStrataGeneprofile\.... Confirm that thermo-cycling program is the same as shown in Fig. 5.4 below.

5.10 (Mx4000) In the "File" menu, select "Import Well Names" followed by "from an Excel File".

5.11 (Mx4000) Select the appropriate Excel file containing the well names from the file N:\Ves Dis Grp\Quality\Archive\Molecular Biology\PCR\Tracking Sheets; select the relevant tracking sheet according to assay (WRL …) number and click "Open".

5.12 (Mx4000) Decline the option to open the "Full screen view" by clicking "No".

5.13 (Mx4000) Ensure the FAM filter channel is displayed and that data will be collected from this channel.

5.14 (Mx4000) Save the run by selecting "Save As" in the "File" menu. Save the file in the "WRL" folder located in the "C:\Program Files\Stratagene\Mx4000\Storage\WRL\...." folder, naming the file with the assay number (i.e. WRL….). This stores the data in a local directory on the attached computer thus avoiding potential problems due to network failures.

5.15 (Mx4000) Click the "Run" tab at the top right of the screen.

5.16 (Mx4000) Click the "Start" button on the pop-up "Run status" box.

5.17 (Mx4000) Check the "Turn lamp off at end of run" box in the pop-up box.

5.18 (Mx4000) The machine should now begin the thermal protocol.

5.19 (Mx4000) Click on the dye buttons towards the bottom left of the screen to view only the dyes being used in the assay. Unused dyes should be "greyed-out".

The following procedure is an outline of the steps required to operate the Mx3005P thermal cycler for one-step RT-PCR of FMDV (and other vesicular disease viruses).

5.1 (Mx3005P) Switch on the Mx3005P at the back of the instrument. It is advisable to do this some time in advance as it takes time for the computer to link-up with the instrument

5.2 (Mx3005P) Switch on the computer and log on

5.3 (Mx3005P) Open the Mx3005P software by double-clicking on the desktop icon

5.4 (Mx3005P) In "New Experiment Options", select "Quantitative PCR (Multiple standards)" and click OK

5.5 (Mx3005P) If the lamp is switched off (red bulb), switch on the lamp to warm-up by clicking on the bulb on the task bar

5.6 (Mx3005P) Place the PCR plate containing the reactions securely into the 96-well block, ensuring that well A1 is located at the far-left corner

5.7 (Mx3005P) In the File menu open the file "WRL profile Mx3005P.mxp" from the N:\Ves Dis Grp\Quality\Archive\Molecular Biology\PCR\WRLStratageneprofile\..... Confirm that thermocycling programme is the same as shown in Fig. 5.4 below.

5.8 (Mx3005P) In the "File" menu, select "Import Well Names" followed by "from an Excel File".

5.9 (Mx3005P) Select the appropriate Excel file containing the well names from the file N:\Ves Dis Grp\Quality\Archive\Molecular Biology\PCR\Tracking Sheets; select the Relevant tracking sheet according to the assay (WRL …) number and click "Open".

5.10 (Mx3005P) Decline the option to open the "Full screen view" by clicking "No".

5.11 (Mx3005P) Ensure the FAM filter channel is displayed so that data will be collected from this channel.

5.12 (Mx3005P) Save the run by selecting "Save As" in the "File" menu. Save the file in the "WRL" folder located in the "Storage" folder, naming the file with the assay number (i.e. WRL….). This stores the data in a local directory on the attached computer thus avoiding potential problems due to network failures.

5.13 (Mx3005P) Click the "Run" tab at the top right of the screen.

5.14 (Mx3005P) Click the "Start" button on the pop-up "Run status" box.

5.15 (Mx3005P) Check the "Turn lamp off at end of run" box in the pop-up box.

5.16 (Mx3005P) The machine should now begin the thermal protocol.

5.17 (Mx3005P) Click on the dye buttons towards the bottom left of the screen to view only the dyes being used in the assay. Unused dyes should be "greyed-out".

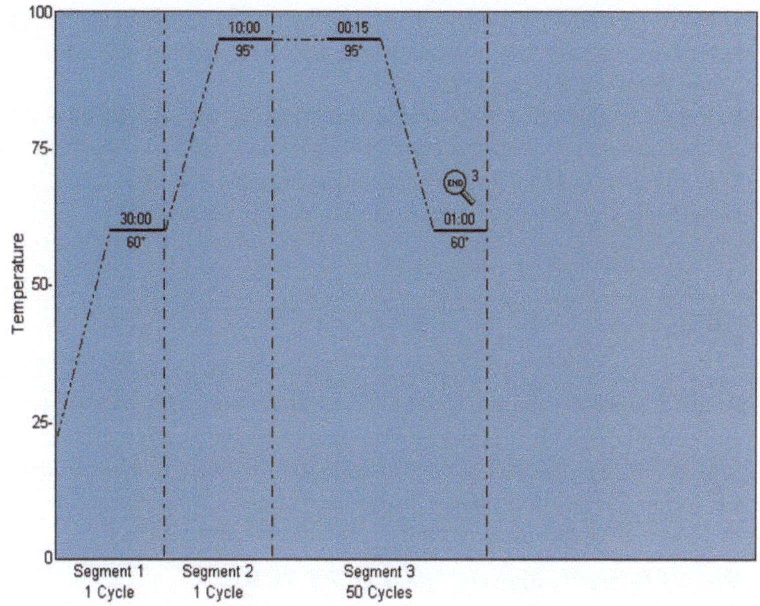

Fig. 5.4 Thermocycling programme used for one-step RT-PCR amplification of vesicular virus RNA

6 Results

6.1 In the "Analysis Selection/setup" module of the Mx4000/MxPro software "Analysis" tab, select the appropriate wells for analysis.

6.2 Click the "Results" tab.

6.3 From the "Area to analyze" option in the top right of the screen, select "Amplification plots". A screen will display the amplification plots for all of the wells that are selected. Double-click on the blue background and change the Y axis scale from "Linear" to "Log". Click "OK".

6.4 Adjust the threshold line using the mouse so that it cuts the amplification plots during the log (straight) phase of the amplification plot.

6.5 Print the amplification plots via the "File" menu.

6.6 Select "Text Report" in the top right of the screen in "Area to analyze".

6.7 From the list of check boxes in the bottom right of the window, uncheck all options with the exception of "Well", "Well Name", "Ct" and "Dye". By clicking on the "Well Name" tab at the top of the column the software will automatically sort the samples making viewing much easier.

6.8 Print the text report via the "File" menu.

6.9 Remove the PCR plate from the instrument by clicking on the "open door" button on the task bar, removing the plate and closing the door using the "close door" button on the task bar.

6.10 Exit the Mx4000/MxPro software (click "Yes" to the save changes option).

6.11 In "My Computer", make a copy of the file and store it in the "N:\Vesicular Dis Grp\quality\ARCHIVE\Molecular biology\pcr\RT-PCR results" folder.

6.12 Log off the computer and switch off the Mx4000/Mx3005P.

7 Analysis

7.1 Positive control samples should have values consistent with those of the daily monitor (within 2 standard deviations of the "current" mean Ct at the time of testing). Negative control wells on the assay plate should give "No Ct" values. The performance of positive control samples can be monitored by reference to the daily monitor records (QAU Form 262).

7.2 **Results Interpretation (for one-step RT-PCR)**
Samples with Ct values of <32.0 are considered positive
Samples with Ct values of "No Ct" are considered negative
Samples giving mean Ct values of ≥32.0–<50 are "inconclusive" and repeat testing (see below) may be required to define a result.

These cut-offs are based on data generated from testing epithelial suspensions [3]. Samples producing Ct values between 32.0 and 50 should be retested (in duplicate) in order to establish a clear positive or negative result. Primary material (prior to RNA extraction step) should be used for this retesting: the samples are designated negative if both wells yield a "No Ct" signal, while a positive result could be reported if both wells generated Ct values that were in close agreement even if they were >32.0. In some circumstances, retesting is not required: such as if the samples are submitted from premises already known to be infected. Final interpretation of these results should be made by the PCR test manager (or deputy) in discussion with the Head of the Vesicular Disease Laboratory.

8 Trouble Shooting

8.1 Positive and negative sample controls are always included within each batch of samples processed in each assay.

8.2 The quantitative PCR results are saved after each PCR amplification (Ct values and amplification plots showing the derived threshold line) so that results can easily be retrieved and re-assessed. Similarly, lists of sample orders can be retrieved from the MagNA Pure LC if required for consultation.

8.3 Technical problems or performance issues with the Mx4000/Mx3005P Sequence Detection System can be discussed with Stratagene (Tel: 0800 917 3281) or email: QPCR@stratagene.com. At the time of writing, firms and contact telephone number(s) were correct. However, these may change so companies currently responsible for maintenance and repair can be found on QAU Form 158.

9 Bibliography

1. Reid, SM, Ferris, NP, Hutchings, GH, Zhang, Z, Belsham, GJ, Alexandersen, S. 2002. Detection of all 7 serotypes of FMDV by real-time, fluorogenic RT-PCR. *J. Virol. Methods*, **105**, 67–80.
2. Reid, SM, Grierson, S, Ferris, NP, Hutchings GH, Alexandersen, S. 2003. Evaluation of automated RT-PCR to accelerate the laboratory diagnosis of FMDV. *J. Virol. Methods*, **107**, 129–139.
3. Shaw, AE, Reid, SM, Ebert, K, Hutchings, GH, Ferris, NP, King, DP. 2007. Implementation of a one-step real-time RT-PCR protocol for diagnosis of foot-and-mouth disease. *J. Virol. Methods,* **143** (1): 81–85.

SOP 27. Differentiation of Sheep and Goat Poxviruses by Real Time PCR

Courtesy of Charles Euloge LAMIEN and Mamadou LELENTA

Animal production Unit,
Animal Production and Health Subprogramme,
FAO/IAEA Agriculture and Biotechnology Laboratory,
Joint FAO/IAEA Division,
IAEA Laboratories
A-2444 Seibersdorf, Austria

Contact email: C.Lamien@iaea.org

1. Purpose

This procedure describes how to detect and differentiate sheep poxvirus from goat poxvirus using Fluorescence Resonance Energy Transfer (FRET) assay with linear probes.

2. Scope

Sheep pox and goat pox are both economically important animal diseases and are included in the list of serious transmissible animal diseases to be notified to the World Organization for Animal Health, the OIE. They are caused by two viruses: goat poxvirus (GTPV) and sheep poxvirus (SPPV). These two viruses are very closely related and together with Lumpy skin disease virus (LSDV), a pathogen of cattle, they form the genus Capripoxvirus within the *Poxvirus* family.GTPV grows readily in both sheep and goat, and SPPV behaves similarly in sheep and goat. However, the pathogenicity of the virus for the host may differ from one species to another according to the virus origin: goat strains being more pathogenic in goat than in sheep and vice versa. However, it is impossible to differentiate clinically a pox disease caused by either GTPV or by SPPV. It is also impossible to differentiate the viruses serologically. However, phylogenetic studies based on partial sequences and complete genomes suggest that they are distinct from each other and from LSDV. Indeed, in our laboratory, a phylogenetic study of capripoxviruses strains based on their chemokine gene sequence has shown that they can be separated into 3 clusters composed of respectively SPPV, GTPV and LSDV strains. The real time PCR described here was designed based in relation to the chemokine gene and was proven to be able to differentiate GTPV from SPPV based on the melting curves in the FRET technology. The limit of sensitivity of the assay is 20 DNA genome copies per PCR reaction.

3. Samples and DNA Extraction

The Qiagen AllPrep DNA/RNA extraction kit is used. Infected tissues are homogenized in PBS using a tissue homogenizer (for example Dispomix, HDV life science). Both homogenized tissues or infected cell culture supernatant are mixed with the extraction kit lysis buffer, RLT plus (200 μL samples + 800 μL RLT plus). The DNA is extracted according to the manufacturer instructions. The extracted samples are eluted in 50 μL of the elution buffer included in the same kit.

4. Real Time PCR

4.1 Adaptation of iCycler to the FRET Chemistry

The iCycler is not designed to have a FRET channel and the machine was adapted for this task requiring a suitable filter positioning for the FRET experiment. A FRET assay needs an excitation of the reporting fluorophore (CY5) by a donor fluorophore (FAM). In the iCycler, we have replaced the Cy5 excitation filter (635/30X) with a FAM excitation filter (490/20X). In the plate set up, the acceptor/reporting fluorophore which is the Cy5 (680/30M) should be selected for the detection of the amplifications.

4.2 Preparation of Well Factor for FRET Experiments

The iCycler needs a fluorophore for calibration: in the Bio-Rad iQ SYBR Green Supermix ROX is included but the iQ Supermix does not contain any dye/fluorophore for calibration. When using the iQ supermix, a so called well factor plate (a plate containing a dye fluorophore for calibration) is needed. Ethidium bromide can be used to prepare the well factor plates.

Ethidium bromide (15 μL) is added (10 mg/mL stock) to 5 mL of 1X PCR buffer to give a final concentration of 300 μg/mL. A well factor plate is then prepared by adding 20 μL/well of this mix.

4.3 Procedure

The following primers and probes are used:

Primers

CpRt forward: 5′-gatagtatcgctaaacaatgg-3′
CpRt reverse: 5′-atccaaaccaccatactaag-3′

Probes

Cp-LNA-FAM1: 5′- acctagcTgtAgttcaCccagtaaa -3′-FAM*
Cp-Cy5: 5′-tcaatttcaataaggacaaaacgatatgga-3′ Phosphate
* The upper characters represent Locked Nucleic Acid.

PCR mix
PCR is carried out in a volume of 20 μL containing the following components:

	Volume (μL)	Final concentration
Primer CpRt forward (5 pmol/μL)	0.4	100 nM
Primer CpRt reverse (5 pmol/μL)	1.2	300 nM
Probe Cp-LNA-FAM1 (5 pmol/μL)	1.2	300 nM
Probe Cp-Cy5 (5 pmol/μL)	1.2	300 nM
iQ Supermix 2×	10	1×
Water	4	
DNA		Variable
Total	18	

Add 2 μL of extracted or positive control DNA to the corresponding tube and 2 μL of water to the No Template Control tube.

Thermal cycling protocol

Cycles	Repeats	Temperature	Time	Remarks
1	1×	95°C	3 min	
2	40×	95°C	10 s	
		46°C	30 s	(data collection (DC))
		72°C	30 s	
3	1×	95°C	30 s	
		40°C	30 s	
4	50	40–90°C	10 s	1°C (Temperature increment) DC

Cycle 2, 46°C: Data collection and real-time analysis (amplification curves) are enabled at the end of the annealing step (46°C).

Cycle 4: Melting analyses

Increase from 40°C by 1°C for 10 s up to 90°C (50 repeats from 40°C).

Melt curve data collection and analysis are enabled at each temperature point. The data obtained are analyzed with the iCycler 3.1 Optical System Software or the BioRad iQ5 2.0 standard Edition Optical System Software.

5. Results

The T_m of the acceptor probe with GTPV is 66°C and that with SPPV sequence is 49°C (Fig. 5.5).

For quantification purpose, a ten-fold serial dilution of plasmids containing the appropriate Capripoxvirus chemokine is prepared and amplified. The standard curve is derived by plotting the cycle threshold (Ct) against the Log copy number. Alternatively, a DNA can be extracted from serial dilution of tissue culture titrated SPPV or GTPV and amplified. The Cts are plotted against the Log $TCID_{50}$ values.

The efficiency calculated on the basis of the serial dilution of plasmids or DNA extracted cell culture titrated virus should be at least 80% and with an R^2 value of at least 0.98.

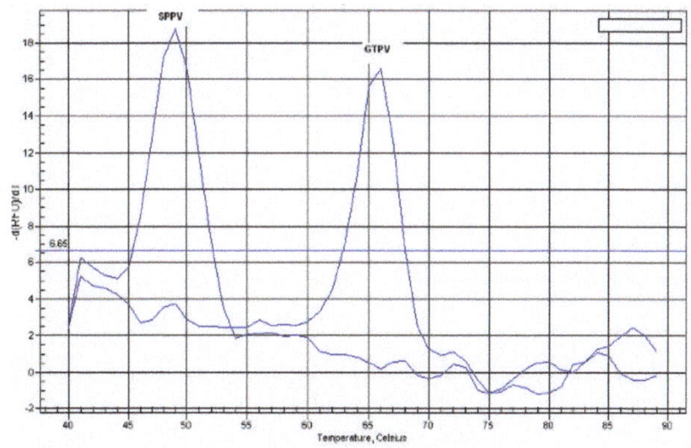

Fig. 5.5 Melting curves of sheep poxvirus and goat poxvirus

6 Appendix

6.1 Material

All DNA/RNA prep (Qiagen)
B-Mercaptoethanol
Ethidium bromide (10 mg/mL), Sigma
iQ Supermix, BioRad (Cat N. 170-8860)
Multiplate PCR plates 96 wells white, BioRad (Cat N. MLP 9651)
PCR sealers microseal B film, BioRad (Cat N. MSB 1001)
PCR water
Probes and primers

Chapter 6
PCR Laboratory Set-up

Contents

As discussed in previous chapters, PCR is a powerful technique for amplifying very small amounts of nucleic acid templates, and also it is known for being a very sensitive technique. Thus, even a few molecules of template DNA can be amplified to billions of copies in a single reaction. Therefore, it is important that only the template we wish to amplify enters in the reaction. Unfortunately, the PCR chemistry has one drawback that is the susceptibility to contamination from its own product and also from external sources. That is why it is important to keep in mind a good laboratory scenario in which the necessary precautions and equipments are set and placed on the right position to avoid unnecessary delays in experiments, and undesirable or false results. This chapter will be dedicated to draw a picture on how a typical PCR laboratory should be handled and set up to obtain reliable and contamination-free results. Also, general guidelines for the establishment and the maintenance of a clean environment will be addressed. Courtesy is given to Viljoen et al, 2005, Molecular Diagnostics PCR Handbook., Springer, 2005.

E.A. Pestana et al., *Early, Rapid and Sensitive Veterinary Molecular Diagnostics - Real Time PCR Applications*, DOI 10.1007/978-90-481-3132-7_6,

6.1 Establishment of a PCR Laboratory

The arrangement or design of a PCR laboratory is of great importance in maintaining a high standard of performance. For that propose, three distinct laboratory areas are required, the reagent preparation area (1); the specimen preparation area (2), and the PCR detection area (3).

The specific technical operations, reagents, and personnel considerations for each of these areas will be discussed below.

6.1.1 Minimum Layout Requirements for a Basic PCR Laboratory

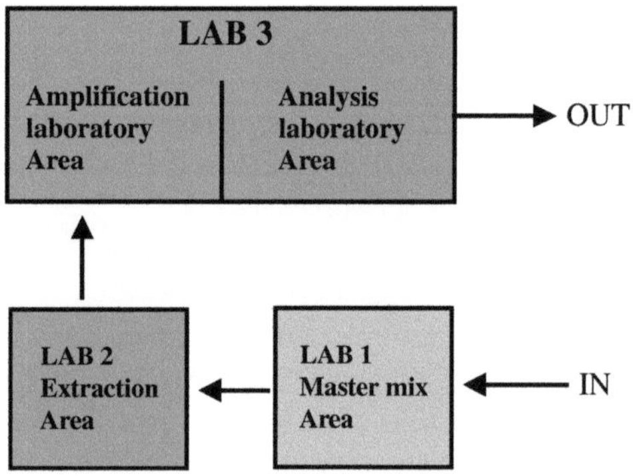

Fig. 6.1 Showing the three distinct laboratory areas required for a basic PCR laboratory set-up

6.1.2 Ideal Physical Arrangement for a PCR Laboratory

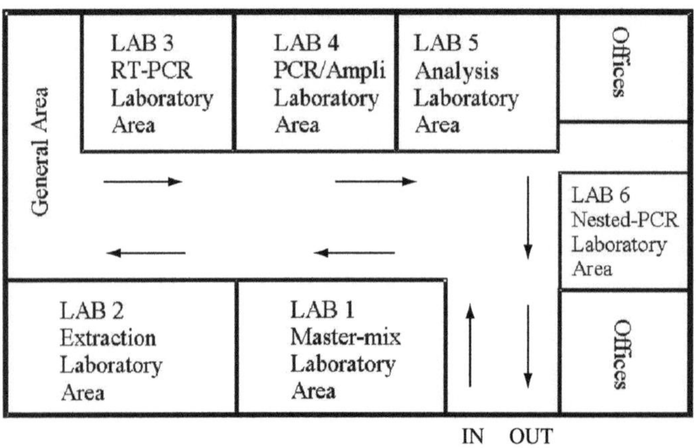

6.1.3 Ideal Physical Arrangement for a Real-Time PCR Laboratory

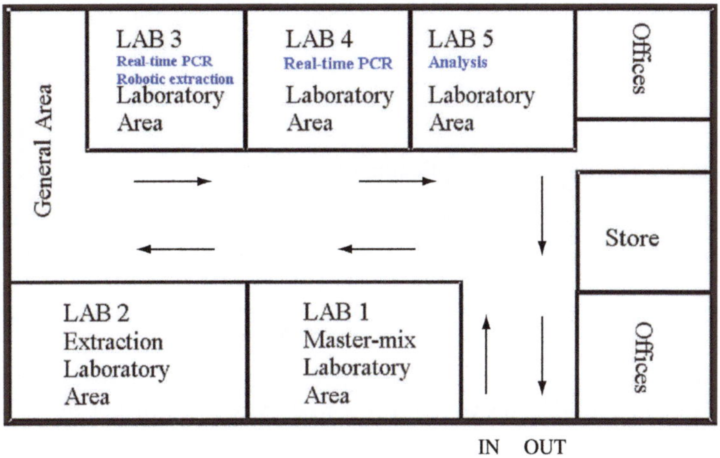

6.1.4 Reagent Preparation – Area 1

Area 1 is devoted to the first steps followed to set up a PCR reaction. In this area, master mix (amplification reagents) reagents are prepared along with the controls.

Prepared master mix and controls are then added to the PCR reaction tubes. Therefore, Area 1 must be maintained clean and free of all sources of contamination.

It is advisable to divide this area into two spaces so as to keep the control DNA away from the master mixes.

6.1.4.1 Equipment Required in Area 1

Each room must have its own separate set of equipment, including pipettors, reagents, pippetor tips, racks and other necessary instrumentation. Each of these should remain on each area, and not be removed from this. If possible equipment should be labelled and used in that location only. Lab coats should also remain dedicated for each area. Pipettors are the most important instrument on PCR preparation and post-handling, therefore, it is important that each pipettor remains devoted to each area, and labelled if possible.

Any standard Area 1 should count on with the following equipment:

- If using one room, a dead-air box with an optional ultraviolet (UV) light or, if using two rooms a clean, dedicated area.
- Freezer/Fridge and ice-boxes.
- Dedicated micropipettes with plugged (aerosol-barrier) tips (25, 50, and 100 μL). or
- Dedicated positive-displacement pipette and tips. A repeat pipettor (optional).

- PCR System reagents (Taq Pol, buffer etc), and consumables (tubes, caps, base, tray, and retainer).
- Microcentrifuge.
- Lab coat.
- Gloves.

6.1.4.2 Important Considerations Applicable to Area 1

- With PCR diagnostics, always *work in a one-way direction* from Area 1 to Area 2 to Area 3 in order to avoid carryover contamination from amplified products.
- Area 1 should be located as far away from Area 3 as possible.
- Area 1 should *not* be in close proximity to the centrifuge area (Area 2) to avoid possible aerosol contamination.
- Clean, dedicated area working surface *must be decontaminated prior* to use with 10% bleach solution (freshly made daily) followed by a 70% ethanol rinse or equivalent.
- If a UV box is used, for your own safety, you *must turn off* the ultraviolet light in the box *before* placing your arms and hands into the working area.
- Master Mix and controls are to be prepared in this area.
- Prepared master mix and controls will be added to PCR reaction tubes and capped in this area.
- Dedicated pipettes with plugged (aerosol-barrier) tips or positive-displacement tips, or a repeat pipettor are required in Area 1 and should *not be transferred* to either Area 2 or Area 3.
- Lab coats designated to wear in Areas 1 and 2 should *never* be worn into Area 3 to avoid carryover contamination of amplified product from Area 3 back to Area 1 or Area 2; disposable coats may be worn over a cloth coat.
- Gloves must be *worn at all times* for your safety as well as for control of contamination from one area to another.
- Gloves are to be changed at each of the three work areas.
- Gloves worn in Area 3 must *never* be worn in Area 1 or Area 2.

6.1.5 DNA/RNA Extraction – Area 2

This area is devoted exclusively to the sample preparation only. Tissue cultures, tissue specimens, and serum samples are all brought into this room and processed for the extraction of RNA and DNA, depending on the application to be pursued. Aerosols are the main problem when processing samples therefore; it is advised to centrifuge samples prior opening the tubes. DNA samples should be manipulated with specialized barrier or positive-displacement pipettes, which prevent the carry-over of aerosols created during pipetting. Personnel protection such as lab coats and gloves must be a prerequisite when handling samples on Area 2.

After the specimens are prepared in this area, they are considered ready to be used in amplification.

6.1.5.1 Equipment Required in Area 2

- Biological cabinet, flow cabinet, dead area space.
- Standard clinical table top centrifuge and/or microcentrifuge (max RCF 16,000*g*).
- Dedicated micropipettes with plugged (aerosol-barrier) tips (25, 50, and 100 μL).
  or
- Dedicated positive-displacement pipette and tips.
- Repeat pipettor (optional).
- Balance/Scale.
- Spectrophotometer.
- Vortex mixer.
- Dry-heat temperature block.
- Sterile calibrated transfer pipettes.
- All reagents and chemicals.
- All consumables.
- Lab coat (different coat to Area 1).
- Gloves (new pair).

6.1.5.2 Important Considerations Applicable to Area 2

- Working surface in cabinet *must be decontaminated prior* to specimen preparation with 10% bleach solution (freshly made daily) followed by a 70% ethanol rinse or equivalent.
- Cabinet *must* be on 30 min *before* specimen preparation procedure.
- All pipettors, pipettes, bulbs, and other equipment used in specimen preparation should be kept in this cabinet at all times.
- Specimens must be stored *separately* from reagents-two separate refrigerators are required.
- Dry baths are preferred over water baths water baths can contaminate specimens through seepage into poorly stoppered tubes.
- A set of pipettes should be *dedicated exclusively* to be used for specimen preparation.
- To avoid specimen-to-specimen contamination, plugged (aerosol-barrier) pipette tips or positive-displacement tips, *must* be used in Area 2.
- Lab coats designated for wear in Areas 2 should *never* be worn into Area 3 or 1 to avoid carryover contamination.
- Gloves must be *worn at all times* for your safety as well as for control of contamination from one area to another. Gloves are to be changed at each of the three work areas.

6.1.6 Amplification (PCR) and Detection – Area 3

Area 3 is the place in which all the final steps take place, that is amplification-ready specimens undergo PCR and the PCR products (amplicons) of this PCR are then detected.

After the PCR reaction is ready, the samples need to be analysed and results needs to be interpreted. Area 3 as mention before should be located as far as possible from area 1. The reason is that the DNA product obtained from previous PCR reactions, arising as micro-aerosols produced during handling of the PCR products, can travel to other Areas (2, 1) if the personnel, or devices from this Area are mixed to another. This area should also be divided into two areas if possible, Amplification area, and Analyses area.

6.1.6.1 Equipment Required in Area 3

- PCR Thermal Cyclers.
- Dedicated micropipettes with plugged (aerosol-barrier) tips (25 and 100 μL) or dedicated positive-displacement pipette and tips.
- Multichannel pipettor, Repeat pipettor, Nonplugged pipette tips (all optional).
- Electrophoreses apparatus and power supplies.
- Microcentrifuge.
- Disposable reagent reservoirs.
- Incubator (37°C ± 2°C) with or without CO_2.
- Microwell plate reader, Microwell plate washer. Key for microwell strip removal.
- UV trans-illuminator.
- Gel documentation system (computer hardware/software & printer).
- Lab coat (different from the one worn in Areas 1 and 2).
- Gloves (new pair).
- Electrophoresis apparatus and documentation system.
- Balance/Scale and microwave oven.

6.1.6.2 Important Considerations Applicable to Area 3

- Area 3 should be kept as far away as possible from Areas 1 and 2 to avoid aerosol contamination.
- Incubator temperature should be kept stable.
- Traffic in and out of incubator should be reduced to a minimum.
- Area 3 pipettes should *never* be used in Areas 1 or 2.
- Pipettes with plugged (aerosol-barrier) tips are used for pipetting denaturing solution into PCR tubes and denatured amplified product into microwell plates.
- Non-plugged tips may be used for all other reagent additions a pipette contaminated with this highly concentrated product could cause false-positive results plugged tips therefore are required to prevent this potential carryover contamination.

- *As this is a one-way workflow*, the lab coat worn in Area 3 must *never* be worn in Areas 1 and 2.
- Gloves must be *worn at all times* for your safety as well as for control of contamination from one area to another. Gloves are to be changed at each of the three work areas. Gloves worn in Area 3 must *never* be worn in Area 1 or Area 2.

6.1.7 Contamination and Sources

As discussed above, PCR contamination is a very big issue when dealing with PCR performance and accuracy. If the test sample has even the smallest contamination of DNA from the target, the reaction could amplify this DNA and report a false positive identification. For example, in an assay containing 1000 samples, each sample must be run with another set of negative controls when contamination is observed. The lack of amplification product within the PCR negative control is not determinative of a positive PCR result in a sample in which contamination is rare. This kind of sporadic contamination is especially problematic in an extremely large throughput assay in which 5–10 negative controls are run for approximately every 1000 samples. Statistically, the likelihood of sporadic contamination in, for example, 1000 samples will not be detected in only 5 negative controls.

Previously, we have discussed the procedures to follow in each laboratory area, in order to avoid in-situ contamination and carryover contamination. In this section we will discuss in more detail possible sources of contamination and also give guidelines how to avoid them.

6.1.7.1 Amplicon Aerosols

Amplicon aerosols are generated by the same PCR product that has been previously amplified. This sort of contamination is usually produced during post-PCR analysis (Area 3). To eliminate aerosol contamination it is advisable to use filter tips as a solution to address this issue. These tips have a porous filter positioned inside the tip. During pipetting, air flows through the filter, which captures aerosols, preventing contact with the pipette's shaft and, subsequently, sample carryover.

6.1.7.2 Template Contaminants

If the PCR is designed to be very sensitive, or if the PCR is used for diagnostics, possible contamination must be eliminated because it can greatly have an impact on the results. Contaminating template or contaminating DNA can come from 3 sources, DNA from other test samples, DNA from experimental materials, such as clones or DNA produced on previous PCR amplifications (carryover contaminants). Therefore, it is advised to always include negative control reactions. Other measures to be considered to control template contaminants are listed below.

- Use of dedicated pipetting devises, and tips containing aerosol barrier.
- After PCR is performed it is recommended to clean the surfaces used i.e. bench cleaning with sodium hypochlorite, 10% bleach solution or 70% Ethanol.
- Preparation of a Master Mix to reduce the number of pipetting steps.
- Always assemble the PCR reaction in a separate Area containing a laminar flow hood with UV light (as discussed above).
- Amplify the PCR with the minimum of cycles required to avoid amplification of non-desire template or false positives.
- Irradiation of the Master Mix prior to the adding of DNA template to be amplified. DNA is sensitive to UV light but primers and dNTPS are more resistance. The logic within this approach is that if there is any source of contamination in the pre Mix, this will be diminished by the UV light.
- Use of DNAse I in the master mix prior to the adding of the target template will have the same effect as the UV light if any DNA contaminant is present.
- The use of the Uracil glycosylase (UDG) decontamination method, which consists on substituting PCR product DNA with uracil (dUTP) bases in place of thymine (dTTP). After amplification, the DNA produced will contain uracil bases. If the enzyme uracil DNA glycosilase (UDG) is added to those products, will result on the removal of the urasil sites, making the PCR product labile and not able to further undergo re-amplification. Here the principle is that if the laboratory routinely uses the uracil incorporated templates, all templates produced will have uracil, but new DNA template to be used will not. Before a new sample is processed, will be first exposed to UDG enzyme. If UDG come across with contaminating U-containing DNA strands from a previous amplification, the U's are cleaved, leaving the strands with gaps. During cycling heating, the strands will fall apart and will not be amplified.
- After PCR it is recommended that samples are centrifuge prior opening to avoid aerosol formation. It is also suggested that a good way to control aerosol is to open the samples when frozen.
- Always wear disposable gloves.

6.1.7.3 Real Time PCR Systems and Contamination

The use of Real Time PCR systems greatly decreases the risk of contamination on the PCR, due to the fact that post-PCR handling is no more required; when the PCR is finished detection and analysis are also completed, and so reaction tubes do not need to be reopen thus eliminating amplicon escape.

The practices mention above should be followed more tightly when using fluorogenic 5' nuclease chemistries (TaqMan) or SYBR® Green I dye chemistry. In the case of SYBR® Green I, potential contamination can be introduced to the samples, and since this dye has non-specificity, any double-stranded DNA will be labelled and detected. Therefore, it is recommended to always check for non-specific product formation by using dissociation curve or gel analysis. Care must be taken to avoid contamination with target DNA, i.e. gene expression assays that span exon-exon junctions minimise the effect of genomic DNA contaminants.

In the case of TaqMan assays, UDG pre-treatment can prevent the re-amplification of carryover PCR products (carryover contamination). There are available master mixes in the market containing the UDG component. The same principle can be applied to SYBR® Green I amplification protocols when suspicion of carryover contamination is present.

6.1.8 Establishment of a PCR Assay

The selection of a PCR assay must be based on factors including scientific and international acceptance, cost, available resources, nature of the intended use, sensitivity and specificity, number of tests to be done and availability of standard reagents.

Development and optimization of the assay must then be performed and should include a series of experimental procedures and the evaluation of the data generated. Analysis should determine a fixed protocol for use, the nature and number of controls required and specifications required of reagents and equipment.

The optimisation of the PCR should be followed by evaluation of the test. This will include a period of testing samples with known histories.

6.1.9 Validation of the Assay

Validation establishes the performance standards of the assay (e.g., sensitivity, specificity, accuracy, precision, positive/negative cut-off criteria etc.), using appropriate statistical methods.

Validation can include a comparison with other methods, with reference standards, collaborative studies with other laboratories using the same test procedure, experimental challenge studies or reproduction of data from accepted standard methods or a reputable publication.

Validation has to main components

- *Specificity* is the ability of the assay to exclusively detect the agent of interest.
- *Sensitivity* is described as the ability to precisely detect small quantities of the test agent.

6.1.9.1 *Validation Pathway or Steps Followed for Assay Validation*

Stage 1

Involves the development of the assay performing a feasibility study to determine whether the assay can detect a range of agents (e.g., virus concentrations, virus serotypes/genotypes), without background activity.

This stage include all aspects of development and optimisation of the test, including the identification of the test, the determination of the target template(s), the sequence determination of the primers and the test conditions and criteria.

Stage 2

Involves the evaluation of the test against a panel of known positive and negative template samples to determine the analytical sensitivity and specificity.

Stage 3

This involves the determination of diagnostic sensitivity and specificity, i.e. involving field samples in 2 parallel tests (classical *vs* PCR).

Diagnostic sensitivity (DS) and specificity (DP)

DS determines the proportion of known infected reference animals testing positive [TP/(TP+FN)], while DP determines the proportion of uninfected reference animals that test negative [TN/(TN+FP)]. (TP = true positive; FN = false negative; TN = true negative; FP = false positive).

DS – how many false negatives you get.

DP – how many false positives you get

Stage 4

Entails the continued monitoring of the validity of assay performance in the field by calculating the predictive value of positive or negative results based on estimates of pathogen prevalence in the target animal population. This can only be done satisfactorily if a diagnostic sensitivity and specificity data (stage 3) are available.

Stage 5

Involves the maintenance of validation criteria using internal quality controls. Frequent monitoring for repeatability and accuracy are needed. The OIE also recommends biannual ring-testing to determine reproducibility between laboratories, although annual testing is also described.

6.2 Quality Assurance Programme or Accreditation

The intention of accreditation or the formal recognition of a quality assurance programme by an independent third party is to ensure that GLP, QA (QC – IQC/EQC) is in place within the diagnostic laboratory. Specific standards are used (e.g. ISO 17025) in accordance with the requirements of the inspecting body (e.g., the South African National Laboratory Accreditation Service).

Requirements will include a quality policy, a quality manual, procedures for evaluating and reviewing new and exciting test methods and protocols, the use of blind quality assurance within and between laboratories, monitoring of equipment, ongoing training, an audit programme, regular review of the quality system and documentation and records of all quality activities and evaluations. Standard operating procedures (SOP's) will be required for not only the test methods used, but also for the calibration and maintenance of instrumentation and equipment used.

6.2.1 Proficiency Testing

Proficiency testing is the means used to determine the capability of a laboratory to perform the assay and effectively detect the agent (internal proficiency testing).

Such testing will also contribute to ensuring that within or between laboratories performing routine diagnostic services, a specific assay is performed according to established international standards (external proficiency testing).

6.2.2 PCR Controls

6.2.2.1 Positive Control Samples

These should include,

- Specimen material with a high pathogen concentration.
- Specimen material with a low concentration (preferably the lowest limit routinely detected, or preferably the lowest amount of detectable target agent present in sample material by conventional means and by reported PCR data.
- Specimen material with expected pathogen concentration in an infected case (not essential).
- Extracted positive DNA/RNA can be used only if the above are not available.

In the case of a proficiency testing process, specimen material spiked with a dilution series of virus (e.g., 10–1 to 10–10) should also be included. Virus titre would have been established by means of a golden standard assay.

6.2.2.2 Negative Control Samples

- A negative control consisting of all reagents except extracted sample DNA or RNA - this should confirm absence of amplicon or agent itself or other cross reactive.
- A negative specimen control sample (allantoic fluid, buffer, tissue), if available, which will undergo nucleic acid extraction to confirm absence of contamination in the extraction procedure.
- Extracted nucleic acid derived from pathogen-free sample material obtained previously (e.g., tissue or other medium normally submitted, including tissue, blood, buffer) - should confirm that no non-specific binding to virus-free specimen DNA/RNA has taken place (eukaryotic and prokaryotic genomic DNA or RNA present in sample). Not essential.
- In the case of RT-PCR, an additional negative control consisting of all reagents as well as the extracted sample nucleic acid, but *without* "reverse transcriptase" included, this should confirm absence of DNA or *amplicon* contaminants in both reagents *and* test material.

Bibliography

1. Cone, WR, Fairfax, MR. 1993. Protocol for ultraviolet irradiation of surfaces to reduce PCR contamination. *PCR Methods Appl.*, **3**, Supplement, S15–17.
2. Dieffenbach, CW, Dveksler, GS. 1993. Setting up a PCR laboratory. *PCR Methods Appl.*, **3**, Supplement, S2–7.
3. Dieffenbach, CW, Dragon, EA, Dveksler, GS. 1995. Setting up a PCR laboratory. In PCR Primer. A Laboratory Manual. pp. 7–16. (Eds. Dieffenbach, CW, Dveksler, GS). Cold Spring Harbor Laboratory Press, New York.
4. Dragon, EA. 1993. Handling reagents in the PCR Laboratory. *PCR Methods Appl.*, **3**, Supplement: S8–9.
5. Dragon, EA, Spadoro, JP, Madej, R. 1993. Quality control of polymerase chain reaction. In Diagnostic Molecular Microbiology: Principles and Applications. pp. 160–8. (Eds. Persing, DH, Smith, TF, Tenover, FC, White, TJ). American Society for Microbiology, Washington D.C.
6. Hartley, JL, Rashtchian, A. 1993. Dealing with contamination: enzymatic control of carryover contamination in PCR. *PCR Methods Appl.*, **3**, Supplement, S10–14.
7. Lee, KW, Liu, BHY. 1982. Experimental study of aerosol filtration by fibrous filters. *Aerosol Sci. Technol.*, **1**, 35–46.
8. Lee, KW, Liu, BHY. 1982. Theoretical study of aerosol filtration by fibrous filters. *Aerosol Sci. Technol.*, **1**, 147–62.
9. Madej, R, Scharf, S. 1990. Basic equipment and supplies. In PCR Protocols: A Guide to Methods and Applications. pp. 455–59. (Eds. Innis, MA, Gelfand, DH, Sninsky, JJ, White, TJ). Academic Press, London.
10. McCreedy, BJ, Callaway, TH. 1993. Laboratory design and work flow. In Diagnostic Molecular Microbiology: Principles and Applications. pp. 149–59. (Eds. Persing, DH, Smith, TF, Tenover, FC, White, TJ). *American Society for Microbiology*, Washington D.C.
11. Mifflin, TE. 2007. Setting up a PCR Laboratory. *Cold Spring Harbour Protocol.*, **1**, 5–14.
12. Orrego, C. 1990. Organizing a laboratory for PCR work. In PCR Protocols: A Guide to Methods and Applications. pp. 447–54. (Eds. Innis, MA, Gelfand, DH, Sninsky, JJ, White, TJ). Academic Press, London.
13. Renoux, A, Boulaud, D. The aerosols: physics and metrology (*Lavoisier Technique et Documentation*, Paris, 1998).
14. Tetzner, R, Dietrich, D, Distler, J. 2007. Control of carry-over contamination for PCR-based DNA methylation quantification using bisulfite treated DNA. *Nucleic Acids Res.*, **35**(1), 4.
15. Viljoen, GJ, Nel, LH, Crowther, J. 2005. Molecular Diagnostics PCR Handbook., Springer.

Chapter 7
Analysis and Troubleshooting

Contents

7.1 How to Design Primers for Real-Time PCR Applications

When designing primers, it is of initial importance to define the target area, and secondly the type of application. The BLAST function from the National Center for Biotechnology Information (NCBI) will help to identify the most suitable gene sequence to be used. There are many software programs; some free on

E.A. Pestana et al., *Early, Rapid and Sensitive Veterinary Molecular Diagnostics - Real Time PCR Applications*, DOI 10.1007/978-90-481-3132-7_7,

websites/pages on the internet, dedicated to primer design and primer optimisation. In this chapter the most important factors that need to be taken into consideration when designing and optimising primers are highlighted.

As a rule of thumb, the following guidelines should be followed when designing primers:

- Primers and probes should be selected in a region with a GC content lower than 80%.
- The amplicon size recommended for real-time PCR applications is between 50 and 150 bp. Do not exceed the amplicon size of 300 bp when designing primers as time efficiency is paramount as it will have a delayed effect on the C_T.

When dealing with cDNA or mRNA amplification procedures – it is advisable to design a primer or probe which crosses one exon junction, thus in the case of genomic DNA contamination the latter will not be amplified. (Fig. 7.1 and Fig. 7.2)

- Primers should be 15–30 mer in length.
- Avoid the presence of secondary structures or primer-dimer formation, as they can interfere with the amplification, particularly in SYBR® green applications.

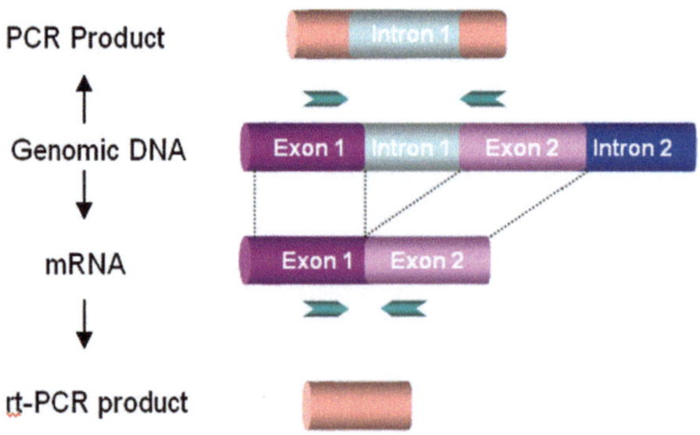

Fig. 7.1 Primers aligning to exons flanking an intron. Any product amplifying gDNA will be much larger than a product amplified from intronless mRNA

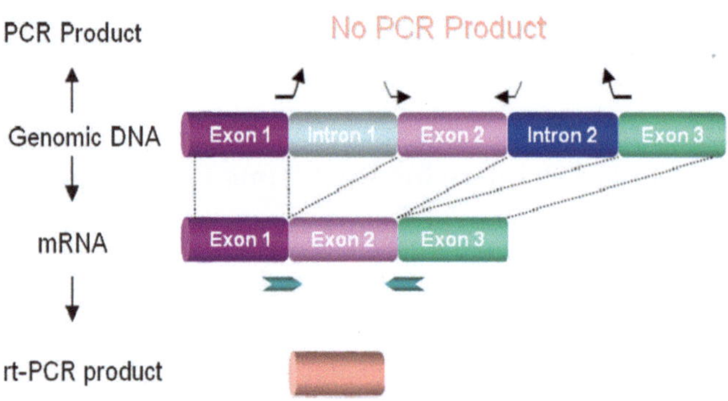

Fig. 7.2 Primers that bridge an exon-exon junction on mRNA. No amplification of gDNA

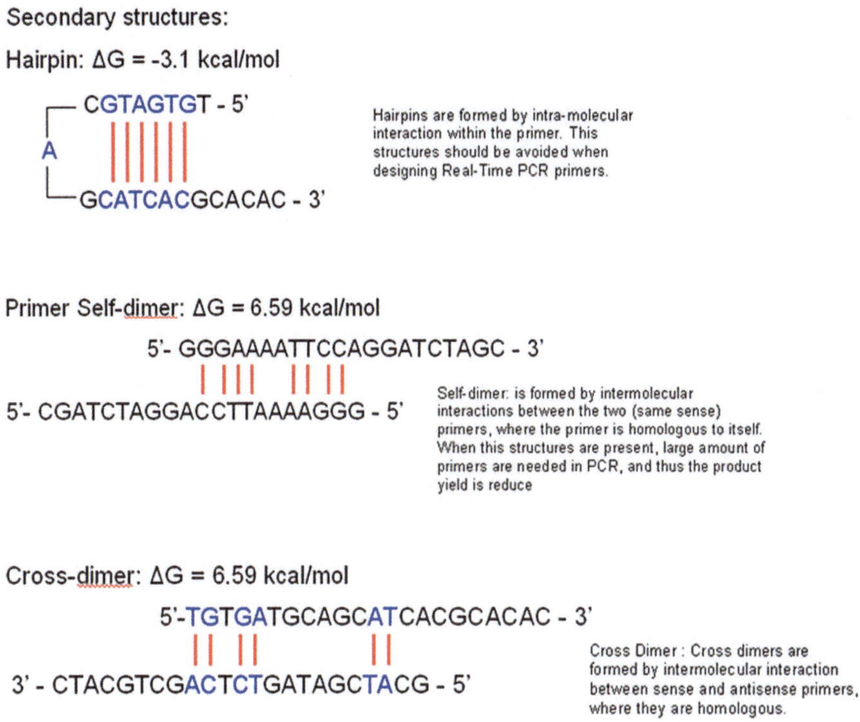

Fig. 7.3 Secondary structures to be avoid during primer design. When Primer-dimer presence is observed, it is advisable to redesign the primers, try to increase the annealing temperature, or try to decrease the primer concentration

The rules stated above are general rules to be followed when designing any PCR primer and are applicable to all chemistries. There are however, specific guidelines and recommendations to be followed for each chemistry or technology application and they will be discussed below.

7.1.1 TaqMan® Probes and Primer Design

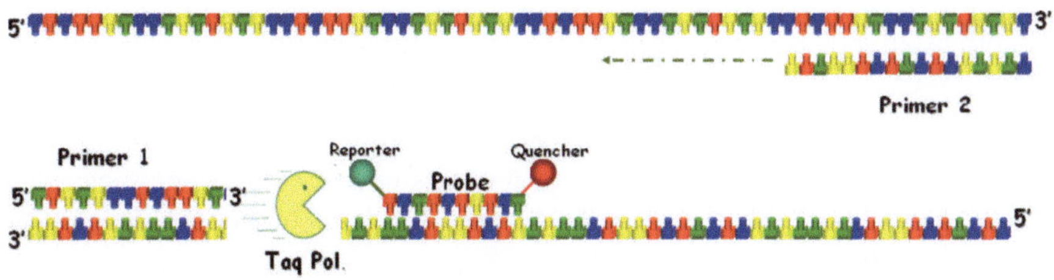

Fig. 7.4 TaqMan probe and primers design and specifications

- The melting temperature (T_m) of the probe should be between 68°C and 70°C. Probes with lower T_m's or with T_m's close to that of the primer-pairs, might not hybridise or bind to the target strand (to be amplified strand), and that can interfere with the intensity of the probe (reporter/quencher) signal.
- The T_m of the probes should be about 10°C higher than that of the primers, to allow for the probe to bind first.
- The primers should be designed as close to the probe as possible. The reason for this is that if the primers are close the probe will be degraded faster and the signal will appear soon after the elongation step has started.
- The G/C contents of the primers should be about 55–60%.
- Primers Forward and Reverse should have the same annealing temperature.
- Runs of identical nucleotides should be avoided, particularly at the 3′ end of the primer. Repetitions should always be avoided for two reasons: (i) possibility of mispriming; (ii) possibility of secondary structure formation.
- Avoid the presence of too many Gs in the 5′ end of the primer.
- Having a G or a C as last nucleotide at the 3′ end may improve specificity.
- Always try to select the strand that contains more Cs than Gs. A higher number of Cs produces a high kinetic hybridization and higher annealing profile.
- For the TaqMan amplification technology, the probe is of paramount importance. So before ordering it, it is advised to test first the primers to be used for the amplification to make sure they will work.
- Please look at *Chapter 3* for all the primer criteria.

7.1.2 Storage of Primers and TaqMan® Probes

Once the primers and probes are reconstituted and/or diluted, it is recommended to distribute them into single-use aliquots and store them at –20°C. Making single-use aliquots limits freeze-thawing and therefore extends their life. It is also important to store TaqMan® probes in the dark to prevent the photo-bleaching which can damage the probe.

7.1.3 SYBR® Green Assays

As discussed in Chapter 3, SYBR® Green chemistry detects all double stranded DNA. When designing primers for this methodology, it is critical to avoid:

- the secondary structures formation mentioned above;
- the non-specific product formation. An excess of free magnesium reduces the Taq polymerase fidelity and may increase the level of nonspecific product amplification.

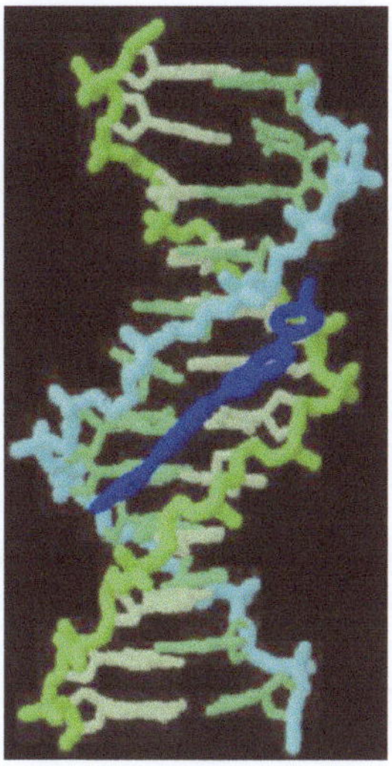

7.1.3.1 SYBR® Green I Behaviour During Real Time PCR

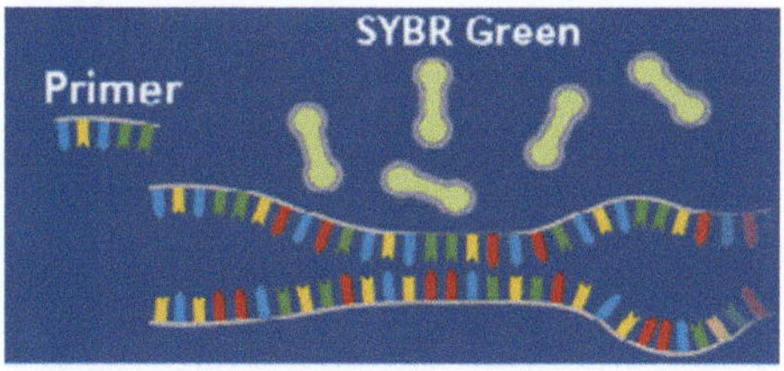

In solution, the unbound dye exhibits very little fluorescence. Fluorescence is greatly enhanced upon DNA-binding. (Courtesy of Roche Diagnostics, GmbH).

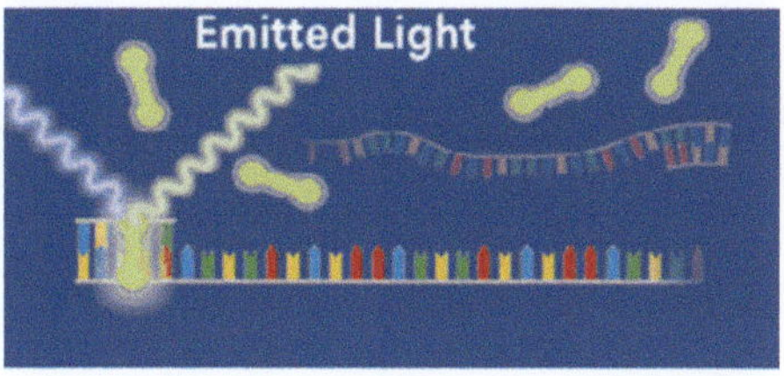

Fluorescent dye SYBR® Green I bind to the minor groove of the DNA double helix. (Courtesy of Roche Diagnostics, GmbH).

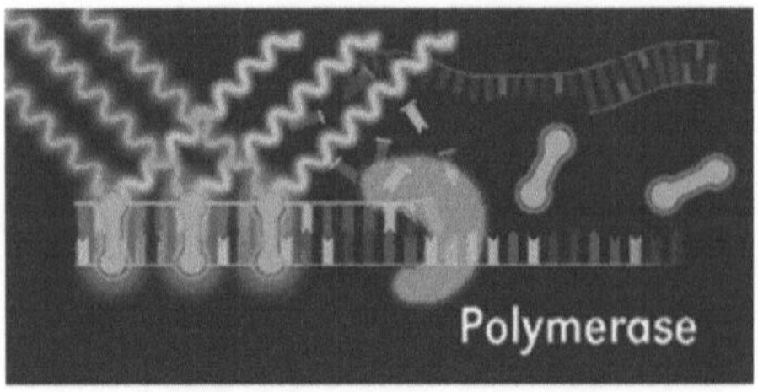

During elongation, more and more dye molecules bind to the newly synthesised DNA. Upon denaturation of the DNA for the next heating cycle, the dye molecules are released and the fluorescence signal falls. Fluorescence measurement at the end of the elongation step of every PCR cycle is performed to monitor the increasing amount of amplified DNA. (Courtesy of Roche Diagnostics, GmbH).

In Fig. 7.5 the graphs show fluorescence vs. the number of cycles for differing initial amounts of template copies. When template concentration is high, amplification of specific DNA is observed during early heating cycles. In case of very few copies of starting template, unspecific products here in green, usually due to primer-dimer formation, can mask the specific signal (blue graph). Here the reduction of magnesium will may help eliminating this unspecific product formation.

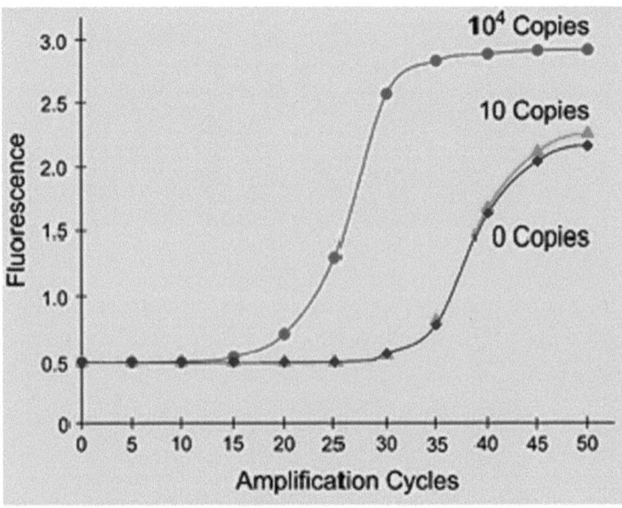

Fig. 7.5 Fluorescence vs. Number of amplification cycles for differing amounts of initial template copies (Courtesy of Roche Diagnostics, GmbH)

7.1.4 Optimisation of Primer Concentration

Primer pairs can be adjusted independently to identify the best T_m and concentration needed to perform the best assay. The concentration of primer in the PCR reaction should be between 0.1 and 0.5 μM. For most PCR applications, including rt-PCR assays and the amplification of longer PCR products, 0.2 μM of each primer produces pleasing results. Primer concentrations which are too high increase the chance

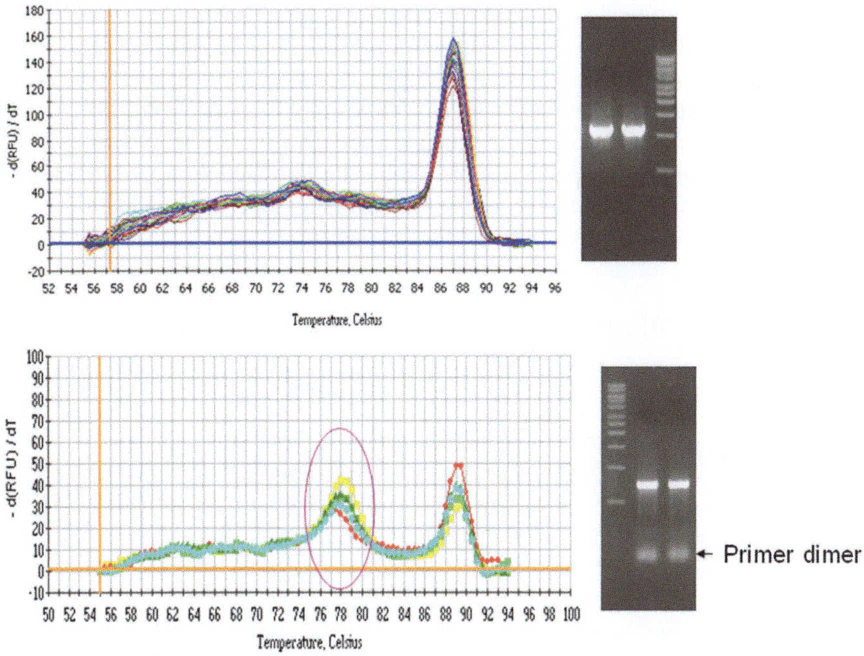

Fig. 7.6 SYBR® Green is a very helpful tool to find out if primers are forming dimer or secondary structures. For that, the amplified product melting curve analysis is carried out as shown in figure. The formation of one single curve will be indicative of the absence of secondary structures, or non-specific products. Primer-dimers will be most prevalent in the No Template Control (NTC) wells and sample wells containing low concentrations of template. In the figures, plots are showing the change in fluorescence as the temperature decreases and the SYBR® Green is freed in the solution. Next to the plots, gel analysis of qPCR products can be observed. (Courtesy of Bio-Rad)

of mispriming. Subsequent extension of misprimed molecules results in nonspecific PCR products.

It is important, especially in Green DNA dye I based real time PCR applications, to minimize the formation of non-specific amplification products. Especially at low target concentrations it is recommended to use the lowest primer concentration without compromising the efficiency of PCR. The optimal concentration of primer pairs is the lowest concentration that results in the lowest C_T and an adequate fluorescence for a given target concentration, with minimal or no formation of primer-dimer. Primers can be efficiently optimised by using SYBR® Green.

7.1.5 Multiple Bands on Gel or Multiple Peaks in the Melting Curve

Agarose gel electrophoresis or melting curve analysis may not always reliably measure PCR specificity. Bimodal melting curves can be sometimes observed in long amplicons (> 200 bp) even when the PCRs are specific. The observed heterogeneity in melting temperature is due to internal sequence in homogeneity (e.g. independent melting blocks of high and low GC content) rather than non-specific amplicon. On the other hand, for short amplicons (<150 bp) very weak and fussy bands migrating ahead of the major specific bands are sometimes observed on agarose gel. These weak bands are super-structured or single-stranded versions of the specific amplicons in equilibrium state and therefore should be considered specific. Although gel electrophoresis or melting curve analysis alone may not be 100% reliable, the combination of both can always reveal PCR specificity.

7.1.6 Effect of Magnesium Chloride and Primer Concentration

Advice

Most laboratories have already standardized set of conditions for rt-PCR. Since what is needed is mostly reproducibility and fastness, most scientists usually do not spend enough time in optimizing their specific PCR. In case of SYBR® Green for example, the best way is to design dilution series of the sample to be analyzed using a standard protocol that works in your laboratory (minor modifications can be used, i.e. annealing temperature or slightly different extension time depending on the length of the amplicon). If the reaction plot is not good, adjustment of cycling parameters, e.g. annealing temperatures, can be made. If after these adjustments the plots still are unsatisfactory, then it is better to order new primers. It is much cheaper than wasting reagents and time trying to optimize the reaction at this point in time.

7.1.7 Molecular Beacons Assays

Low signal-to-background ratio: poor stem formation. In that case, the assay medium may contain insufficient salt. 1 mM $MgCl_2$ in the solution, usually ensures the stem formation. The molecular beacon may fold into an alternate conformation, which results in a sub-population that is not quenched well. To solve this problem it is advised to change the stem sequence (and probe sequence, if necessary).

Incomplete restoration of fluorescence at low temperatures. If the stem of a molecular beacon is too strong, at low temperatures it may remain closed while the probe is bound to the target. This may happen inadvertently if the probe sequence forms a hairpin that results in a stem longer and stronger than originally designed. Change the sequence at the edges of the probe and the stem sequence to avoid this problem.

7.2 Assay Performance Evaluation Using Standard Curves

Theoretically, during the exponential growth phase, there is a quantitative relationship between the amount of the starting target sequence and amount of PCR product at any given cycle. This relationship can be measured by the use of the following formula:

$$X_n = X_0(1+E)^n$$

Where X_n is the amount of the target sequence at cycle n, X_0 is the initial amount of the target, and E is the amplification efficiency. The E values are included between 0 (no amplification) and 1 (every amplicon is replicated every cycle).The emission of fluorescence either by the DNA binding dyes or by the probes in the amplification mix allow the construction of the amplification curves by plotting the increase in fluorescence versus the cycle number. Every amplification curve has three phases: an exponential phase, a non-exponential phase and a plateau. With real-time PCR, the quantitative information is deducted from the exponential phase of the amplification curve.

7.2.1 Threshold Selection

The Threshold line is the level of detection or the point at which a reaction attains a fluorescent intensity above the background. The threshold line is usually set in the exponential phase of the amplification.

The threshold cycle (CT) indicates the fractional cycle number at which the amount of amplified copies reaches a fixed threshold. In simple words, CT is defined as the first cycle in which there is a significant increase in fluorescence above the background. At this point the amount of product is defined as:

$$X_T = X_0(1+E_X)C_{T,X}$$

Where X_T is the threshold number of copies, C_T,X is the threshold cycle, X_0 the intial number of template copies, and Ex the efficiency of target amplification.

A similar equation for the endogenous reference, a housekeeping gene for example, is given by the following equation; to calculate the amount of amplified sample in the endogenous reference control reaction at its Threshold cycle:

$$R_T = R_0(1+E_R)^{C_{T,R}}$$

Where R is denoted as the quantity corresponding to the endogenous reference gene at its Threshold cycle, Ro is the initial number of copies of the endogenous reference gene, ER is the efficiency of amplification of the endogenous reference gene, and $C_{T,R}$ is the threshold cycle number for the endogenous reference gene, where the amplified reference exceed the threshold value.

7.2.2 Quantification of Gene Targets with the Quantitative Real Time PCR: Absolute and Relative Gene Quantification

7.2.2.1 Absolute Quantification (Standard Curve Method)

This method calculates the copy number of a gene by relating the PCR signal to a standard curve.It uses DNA Standards of known quantity for the quantification of the unknown samples. The standards are basically the same target sequences (e.g. Plasmid DNA or purified amplification products) amplified in a dilution series.

The CT values for the unknown samples will be calculated from the correlation curve of the standards. This method is very error prone since the results are very much relying on the technical quality of the standard curves. Pipetting errors for the samples and the dilution series will cause very meaning full errors in quantification using this method. Common real-time PCR applications that employ absolute quantification include chromosome or gene copy number determination and viral load determination.

7.2.3 Relative Quantification

Relative quantification estimates the change in mRNA expression levels by comparing the PCR signal of the target transcript to that of a control (a so-called housekeeping or reference gene).Various mathematical models exist to carry out relative quantification, and some have been incorporated into different software packages for data analysis (e.g. REST, GeNorm, qBasePlus).

Basically a mathematical quantification model transforms Ct values into normalized relative quantities. The first model [10] assumes optimal (100%) PCR efficiency for both the target of interest and the reference targets. But PCR efficiency is not always optimal and the PCR efficiencies of the target and reference are not always equal. Michael Pfaffl improved the model by correcting for differences in PCR efficiency. The most advanced approach from Jo Vandesompele additionally incorporates more than one reference gene for more accurate normalization.

7.2.3.1 Livak Method $2^{-\Delta\Delta CT}$ Method

Unlike the Standard Curve method, the Delta-Delta CT method is an approximation of transcription based on the change in threshold values for control vs. target cells (the cycle number at which fluorescent signaling crosses the "threshold" of logarithmic increases in cDNA concentration.) This method assumes that either there is a minimal correction for the reference gene or reference and target genes have similar efficiencies. Dilution curves are used to ensure that the DCts do not change in order to validate these assumptions. This method is a convenient way to analyze the relative changes in gene expression from real-time quantitative PCR experiments. A drawback of this method is that it assumes $\approx$100% efficient target and standard gene PCR reactions which is usually not the case.

7.2.3.2 Pfaffl Method

Like the Standard Curve Method, this method provides a means for quantification of a target gene transcript in comparison to a reference gene. The relative expression ratio is calculated only from the real-time PCR efficiencies and the crossing point deviation of an unknown sample versus a control. Unlike the Standard Curve Method this model needs no calibration curve, as control levels are included within the model. High accuracy and reproducibility (<2.5% variation) may be reached using this method of data analysis.

7.2.3.3 Vandesompele Method

Since housekeeping gene expression has been reported to vary considerably, Vandesompele outlined a robust and innovative strategy to identify the most stably expressed control genes in a given set of tissues, and to determine the minimum number of genes required to calculate a reliable normalization factor. The geometric mean of multiple carefully selected housekeeping genes was validated as an accurate normalization for accurate RT-PCR expression profiling, which, among other things, opens up the possibility of studying the biological relevance of small expression differences.

7.3 Most Common Problems When Performing Real-Time PCR

7.3.1 PCR Amplification Problems

Little or no PCR product. Poor quality of PCR templates, primers, or reagents may lead to PCR failures. Inclusion of the appropriate PCR controls to eliminate these possibilities is essential. In the case of mRNA detection, it should be born in mind that some genes are expressed transiently or only in certain tissues. The PCR result being dependant of the gene level expression, it is advised to review the literature for the gene expression patterns.

Primer-dimer. Primer-dimer may be occasionally observed if the quantity of target nucleic acid is very low in the starting material. If this is the case, increasing the template amount may help to eliminate the primer-dimer formation.

Non-specific amplicons. Non-specific amplicons give false positive real-time PCR results. They can be identified by both gel electrophoresis and melting curve analysis. To avoid this problem, please make sure to perform hot-start PCR and use at least 60°C as annealing temperature. However, not all hot-start systems are equally efficient at suppressing polymerase activity during the sample setup. If the non-specific amplicon is persistent, you have to choose a different primer pair for the amplification of the gene of interest.

7.3.2 Control Samples

In any diagnostic test, negative and positive controls are essential.

7.3.2.1 Non Template Control (NTC)

Primer-dimer presence in the reaction can also be confused with contamination. To probe the presence of primer-dimers, especially when SYBR® green is not used, a serial dilution of the template followed by an electrophoretic analysis of the final PCR products can help identifying the primer-dimers or the contaminant. In a typical reaction, the higher the concentration of the sample input in PCR is, the lower will be the intensity of the primer-dimer formation. If it is proven that primer-dimer formation is not present but an increase of signal is still observed in a reaction without template, this means that the reaction has been contaminated. In that case, it is advisable to dispose of all the reagents and to replace them with new ones. It is advised to aliquot the reaction components in the DNA free "master mix" using filter tips at all times to avoid carry over contamination.

7.3.2.2 Positive Controls

When performing Quantitative Reverse Transcription PCR (qRT-PCR), two positive controls should always be included in the reaction in order to measure both the efficiency of the reverse transcription (RT) step and the PCR amplification step. When no fluorescence is detected in the RT-PCR control, but the PCR amplification control emits fluorescence, this indicates the failure of the RT step. The same occurs when no fluorescence is encountered in the PCR amplification step but it is present in the RT step – that would be an indication of failure in the PCR step.

7.3.3 Signal Problems in Real Time PCR

1. *Poor amplification of the Target.* Poor amplification of the target can be due to two reasons: poor quality of the template or poor reverse transcription reaction in the case of realtime RT-PCR. It is always important to follow standard procedures to ensure that the sample is as pure as possible (260/280 ratio at around 1.8 for a DNA) and without inhibitors. In the case of RNA as starting material, including a control RNA in the test will give an indication in the efficiency of the reverse transcription (RT-PCR) step. It is important to make sure that the quality of the input RNA is good or that the concentration of the RNA used to generate cDNA is not too concentrated or too diluted. It is also important to make sure that the RT-PCR conditions used are optimal.

2. *Poor amplification of Target when using TaqMan probes.* This can be due to no annealing of the probe to the target DNA. This may happens if the annealing temperature is too high. Therefore it is advisable to recalculate the T_m of the probe.
3. *Amplification of the Target in the NTC.* This result indicates the contamination of the reagents and they should be discarded and replaced by new ones.
4. *Poor PCR amplification efficiency.* The accuracy of real-time PCR is highly dependent on the PCR efficiency. A reasonable efficiency should be at least 80%. Poor primer quality is the leading cause for poor PCR efficiency. In this case, the PCR amplification curve usually reaches the plateau early and the final fluorescence intensity is significantly lower than that of most other PCRs. This problem may be solved with re-designed primers. Standard curve: poor slopes. When performing Real-Time PCR, a slope value of –3.32, is equivalent to 100% efficiency. Poor slope can be due to several reasons, particularly: (i) incorrect dilutions, (ii) presence of inhibitors in the reaction and (iii) improper reaction conditions.
5. *Standard curve: Bad correlation coefficient.* This can be due to: (i) incorrect baseline and threshold settings, (ii) improper pipetting or (iii) incorrect dilutions.

7.3.4 Amplification Plots

A good amplification plot should contain all the PCR 3 phases: the exponential phase, the non-exponential phase and the plateau or end-point phase.

1. *Signal from replicates and control groups do not exhibit a uniform profile.* this can be a result of contamination. The samples should be re-made and the blocks cleaned.
2. *Absence of growth in the amplification plots.* This could indicate (i) a pipetting error, or (ii) the absence of input template in the well, or (iii) an error in the PCR mix.
3. *Well-to-well variance.* Variance between samples can be due to the poor quality of the optical sealing material. Optical clear sails for PCR plates or optical clear cups for sealing of the tubes specifically for fluorescence applications should be used. Well-to-well variance can be seen if the PCR machine is not calibrated.
4. *High cycle threshold (C_T).* High cycle threshold is an indicator of poor PCR efficiency. Many reasons can be at the origin of this poor efficiency:

   - Sample evaporation. Always close the seal in a proper and tight way, and if possible avoid the use of corner wells. Always use good quality seal material as this might interfere with the fluorescence uptake.
   - Primer-dimer formation. Always check the PCR amplicons on a gel, or by using the SYBR® green melting curve to make sure that there is no primer-dimers presence. In case they are present, it advised to redesign the new primers.

- Wrong annealing temperature. The annealing temperatures of primers and probes should be as close as possible to optimum.
- Primer concentration: Use the recommended primer concentrations as discussed in Chapter 3 that would be suitable for your particular chemistry.

7.4 Summary: Optimised Real-Time PCR Assay

Theory given in this and previous chapters will help to determine if your real-time PCR assay is optimised and ready to be applied in your experiments. When evaluating a new real-time PCR assay, it is advised to make a test with known template: prepare a simple dilution series of a known template concentration and compare this with your unknown sample concentration. The standard curves which are generated in this test will help determining the reliability and robustness of the new assay. Standard curves are constructed by plotting the log of the starting concentration of unknown template against the C_T value obtained during the amplification of each of your dilutions of the known template. When using a correct dilution factor, the spacing of the fluorescence curves will be doubled with each amplification cycle.

There are 3 main parameters to be considered when assessing the performance of a realtime PCR assay:

1. *The linear standard curve: correlation coefficient (r) and coefficient of determination (R^2):*
These two values represent the linearity of the experimental data, and thus measure the variability across the replicates, and whether the amplification efficiency is the same in the different dilution series (different starting template concentrations). Slight differences in the C_T values between replicates will lower r or R^2 values. A good standardised assay will normally produce an r whose absolute value is >0.990 or an R^2 value >0.980.

2. *Amplification efficiency (E):*
Amplification efficiency is calculated by using the formula: $E = 10^{-1/\text{slope}}$

Ideally, a PCR reaction in the exponential phase will double its amplicon amount at each cycle, i.e. $E = 2$ for the reaction. Using efficiency equal to 2 in the equation above, $2 = 10^{-1/\text{slope}}$, the optimal slope of the standard curve will be –3.32. In most cases the efficiency of the reaction is represented in percentage and it is calculated by using the following formula:

$$\%\text{Efficiency} = (E - 1) \times 100\%$$

An efficiency close to 100% is a good indicator of a reliable, reproducible and robust assay. In practice good % E values vary between 90% and 105%. Usually low efficiencies are caused by poor primer design or poorly optimised reaction conditions. Higher % E values (>105%) would be an indicator of (i) pipetting errors when performing the dilution series, (ii) or co-amplification of non-specific products (iii) or presence of primer-dimers.

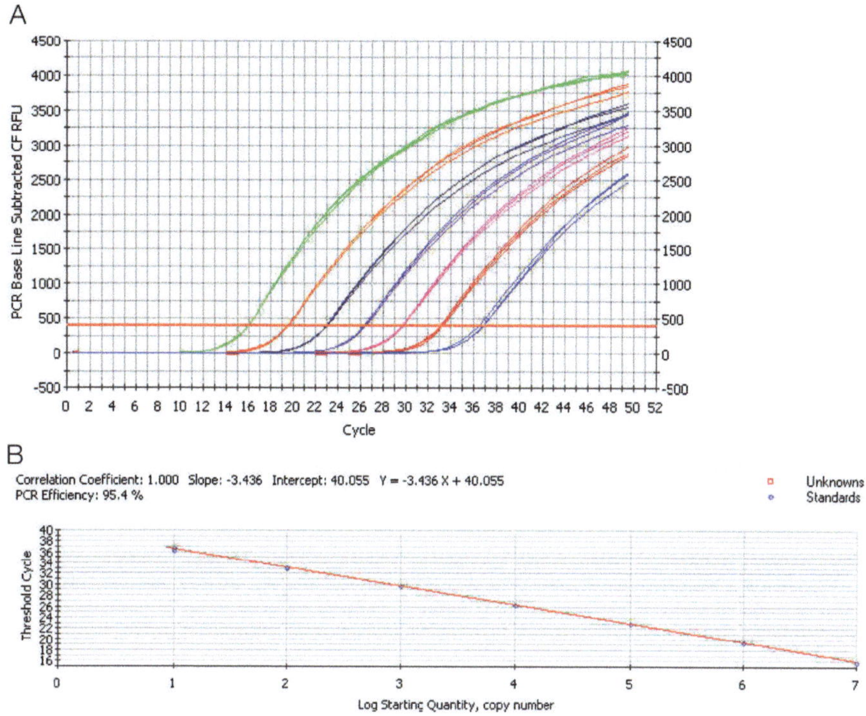

Fig. 7.7 Generating a standard curve to assess reaction optimization. A standard curve was generated using a 10-fold dilution of a template amplified on the iCycler iQ® real-time system. Each dilution was assayed in triplicate. **A**, Amplification curves of the dilution series. **B**, Standard curve with the C_T plotted against the log of the starting quantity of template for each dilution. The equation for the regression line and the r value are shown above the graph. The calculated amplification efficiency was 95.4%. (Courtesy of Bio-Rad)

Presence of inhibitors in the reaction could be reflected by an increase in the efficiency values – by causing a delay in the C_T figure, decreasing the absolute value of the slope, and therefore the calculated efficiency increases (Fig. 7.7).

3. *Replicates:*

It is recommended the use of triplicates when performing Real Time PCR, and the results obtained should be consistent. Representation of evenly distributed amplification curves, and linear standard curve are showed in Fig. 7.7.

Bibliography

1. Hellemans, J, et al. 2007. qBase relative quantification framework and software for management and automated analysis of real-time quantitative PCR data. *Genome Biol.*, **8**(2), R19.
2. Vandesompele, J, De Preter, K, Pattyn, F, Poppe, B, Van Roy, N, De Paepe, A, Speleman, F. 2002. Accurate normalization of real-time quantitative RT-PCR data by geometric averaging of multiple internal control genes. *Genome Biol.*, **3**(7), research0034.1–research0034.11

3. Higuchi, R, Fockler, C, Dollinger, G, Watson, R. 1993. Kinetic PCR analysis: real-time monitoring of DNA amplification reactions. *Biotechnology*, **11**, 1026–30.
4. Morrison, TB, Weis, JJ, Wittwer, CT.1998. Quantification of low-copy transcripts by continuous SYBR® Green I monitoring during amplification. *BioTechniques*, **24**, 954–62.
5. Malinen, E, Kassinen, A, Rinttila, T, Palva, A. 2003. Comparison of real-time PCR with SYBR® Green I or 5′-nuclease assays and dot-blot hybridization with rDNA-targeted oligonucleotide probes in quantification of selected faecal bacteria. *Microbiology*, **149**, 269–77.
6. Bustin, SA. 2000. Absolute quantification of mRNA using real-time reverse transcription polymerase chain reaction assays. *J. Mol. Endocrinol.*, **25**, 169–93.
7. Vandesompele, J, De Paepe, A, Speleman, F. 2002. Elimination of primer-dimer artifacts and genomic coamplification using a two-step SYBR® green I real-time RT-PCR. *Anal. Biochem.*, **303**, 95–8.
8. Souaze, F, Ntodou-Thome, A, Tran, CY, Rostene, W, Forgez, P. 1996. Quantitative RT-PCR: limits and accuracy. *BioTechniques*, **21**,280–5.
9. Giulietti, A, Overbergh, L, Valckx, D, Decallonne, B, Bouillon, R, Mathieu, C. 2001. An overview of real-time quantitative PCR: applications to quantify cytokine gene expression. *Methods*, **25**, 386–401.
10. Livak, KJ, Schmittgen, TD. 2001. Analysis of relative gene expression data using real-time quantitative PCR and the 2(-Delta Delta C(T)) method. *Methods*, **25**, 402–8.
11. Marino, JH, Cook, P, Miller, KS. 2003. Accurate and statistically verified quantification of relative mRNA abundances using SYBR® Green I and real-time RT-PCR. *J. Immunol. Methods*, **283**, 291–306.
12. Pfaffl, MW. 2001. A new mathematical model for relative quantification in real-time RT-PCR. *Nucleic Acids Res.*, **29**, 45.
13. Freeman, W-M, Walker, SJ, Vrana, KE. 1999. Quantitative RT-PCR: pitfalls and potential. *BioTechniques*, **26**, 112–15.
14. Peirson, SN, Butler, JN, Foster, RG. 2003. Experimental validation of novel and conventional approaches to quantitative real-time PCR data analysis. *Nucleic Acids Res*. **31**, 73.
15. Klein, D. 2002. Quantification using real-time PCR technology: applications and limitations. *Trends Mol. Med.*, **8**, 257–60.
16. Pfaffl, MW, Horgan, GW, Dempfle, L. 2002. Relative expression software tool (REST) for group-wise comparison and statistical analysis of relative expression results in real-time PCR. *Nucleic Acids Res.*, **30**, 36.
17. Ririe, KM, Rasmussen, RP, Wittwer, CT. 1997. Product differentiation by analysis of DNA melting curves during the polymerase chain reaction. *Anal. Biochem.*, **245**, 154–60.
18. Lekanne Deprez, RH, Fijnvandraat, AC, Ruijter, JM, Moorman, AF. 2002. Sensitivity and accuracy of quantitative real-time polymerase chain reaction using SYBR® green I depends on cDNA synthesis conditions. *Anal. Biochem.*, **307**, 63–69.
19. Wu, DY, Ugozzoli, L, Pal, BK, Qian, J, Wallace, RB. 1991. The effect of temperature and oligonucleotide primer length on the specificity and efficiency of amplification by the polymerase chain reaction. *DNA Cell Biol.*, **10**, 233–8.
20. Tichopad, A, Didier, A, Pfaffl, MW. 2004. Inhibition of real-time RT-PCR quantification due to tissue-specific contaminants. *Mol. Cell Probes*, **18**, 45–50.
21. Yuan, JS, Feng Chen, AR, Stewart, CN. 2006. Statistical analysis of real-time PCR data. *BMC Bioinformatics*, **7**, 85.
22. Sabek, O, Dorak, MT, Kotb, M, Gaber, AO, Gaber, L. 2002. Quantitative detection of T-cell activation markers by real-time PCR in renal transplant rejection and correlation with histopathologic evaluation. *Transplantation*, **74**(5), 701–7
23. Troubleshooting
24. http://www3.appliedbiosystems.com/cms/groups/mcb_support/documents/generaldocuments/cms_042997.pdf
25. http://www.abgene.com/downloads/Guide_QPCR-probe.pdf

26. http://www.protocol-online.org/prot/Molecular_Biology/PCR/Real-Time_PCR/
27. http://www.ambion.com/techlib/tn/102/17.html
28. http://www.appliedbiosystems.com/support/tutorials/pdf/data_analysis_7700.pdf

Chapter 8
Specifications for PCR Machines

Note: SYBR is a trademark of Invitrogen Corporation. Bio-Rad Laboratories, Inc. is licensed by Invitrogen Corporation to sell reagents containing SYBR Green I for use in real-time PCR, for research purposes only.

E.A. Pestana et al., *Early, Rapid and Sensitive Veterinary Molecular Diagnostics - Real Time PCR Applications*, DOI 10.1007/978-90-481-3132-7_8,

Maufacturer	Bio-RAD		
Model	MyiQ™ (V2.0)	iQ™5 (V2.0)	DNA engine opticon® 2
Appearance	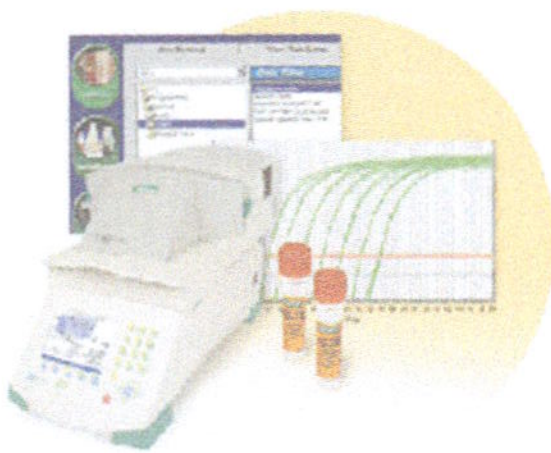	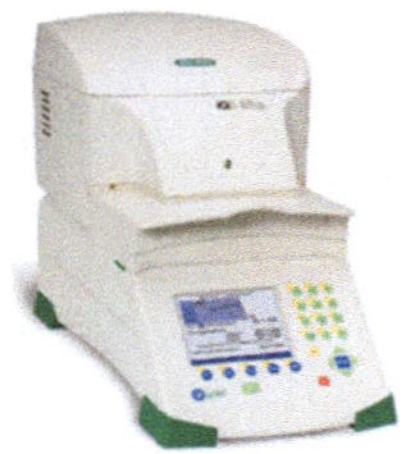	
Detector	CCD	CCD	PMT
Light source	Tungsten-halogen lamp	Tungsten-halogen lamp	96 LEDs
Excitation channels	1	5	1
Emission (detection) channels	1	5	2
Targets	1	Up to 5	1 or 2
Interchangeable filters	X	√ (can be changed on site by customer)	X
Multiple reference gene normalisation	√ – Gene Expression macro for more accurate quantitation	√ – Gene expression macro for more accurate quantitation	√ – Gene expression macro for more accurate quantitation
Ramp rate	3.3°C/s heating and 2.0°C/s cooling	3.3°C/s heating and 2.0°C/s cooling	Up to 3°C/s
Adjustable ramp rate	√	√	√
Heating/cooling method	Peltier block	Peltier block	Peltier block
Gradient function	√	√	√

Maufacturer	Bio-RAD		
Model	MyiQ™ (V2.0)	iQ™5 (V2.0)	DNA engine opticon® 2
Sample format	96-well plate	96-well plate	96-well plate or 12 × 0.2 mL 8-tube
Sample volume range	15–100 μL	15–100 μL	10–100 μL
Detection method	CCD camera (highest sensitivity/accuracy)	CCD Camera+intensifier (highest sensitivity/accuracy)	2 Photomultiplier tubes (PMT)
Data collection	Simultaneous (data collected from all wells at same time)	Simultaneous (data collected from all wells at same time)	96-well LED array illuminated 1 well at a time (3 s for 96-well plate read)
ROX reference dye needed	X	X	X
Open platform chemistry	√	√	√
Dimensions (W × D × H) Model	31.1 × 54.6 × 34.3 cm	31.1 × 54.6 × 34.3 cm	34 × 47 × 60 cm
Weight	17.6 Kg	17.6 Kg	29 kg
Software	MyiQ/iQ5 Soft 2.0 (SE) Comparison of up to 5000 samples	MyiQ/iQ5 Soft 2.0 (SE) Comparison of up to 5000 samples	Opticon monitor™ 3.1
Additional features	"All Candidates" functionality	"All Candidates" functionality	
Heating/cooling method/accuracy/uniformity			

Maufacturer	Bio-RAD		
Model	Chromo 4™	MiniOpticon™	CFX96™
Appearance	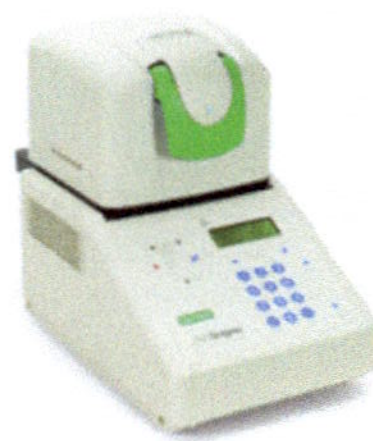	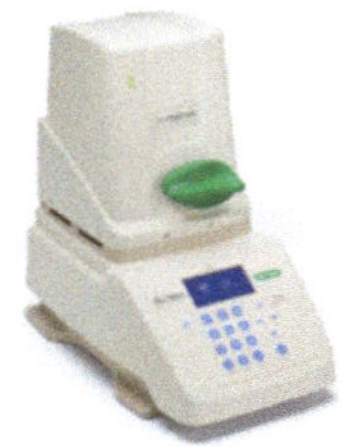	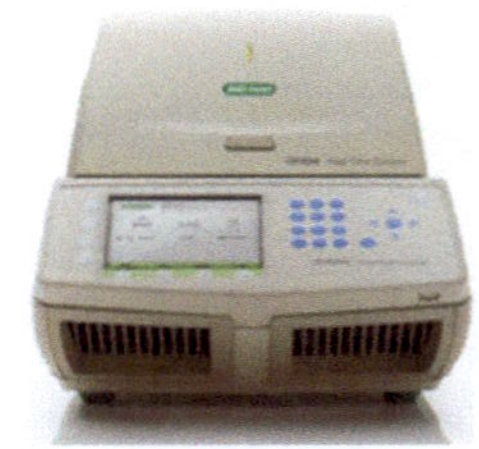
Detector	Filtered photodiodes	Filtered photodiodes	Filtered photodiodes
Light source	4 colored LEDs	48-well LED array	6 colored LED shuttle
Excitation channels	4	1	6
Emission (detection) channels	4	2	6
Targets	Up to 4	Up to 2	Up to 5, plus 1 FRET channel
Interchangeable filters	√ (can be changed by customer)	X	X
Multiple reference gene normalisation	√ – Gene expression macro for more accurate quantitation	√ – Gene expression macro for more accurate quantitation	√ – CFX manager™ software and Gene expression macro for more accurate quantitation
Ramp rate	Up to 3°C/s	Up to 2.5°C/s	Up to 5.0°C/s /av 3.3°C/s "Reduced mass" patent pending Peltier block
Adjustable ramp rate	√	√	√
Heating/cooling method	Peltier block	Peltier block	Peltier block
Gradient function	Ö	√	√
Sample format	96-well plate or 12 × 0.2 mL 8-tube	48-well plate or 12 × 0.2 mL 8-tube	99-well plate or 12 × 0.2 mL 8-tube, regular and low profile vessels
Sample volume range	10–100 mL	10–100 μL	10–50 μL

Maufacturer	Bio-RAD		
Model	Chromo 4™	MiniOpticon™	CFX96™
Detection Method	Photodiodes	Photodiodes	LED shuttle, 96-well detection (12 s for 5 color, 3 s fast scan for 1 color SYBR®/FRET channel)
Data collection	Scanning 4 LED Photonic Shuttle sequentiell illumination of 4 wells at a time	48-well sequentiell LED array fire illuminate 1 well at a time (9 s for 2color 48-well plate read)	Hardware and software normalization. No passive reference needed
ROX reference dye needed	X	X	X
Open platform Chemistry	√	√	√
Dimensions (W × D × H)	24 × 38 × 34 cm with DNA engine (PTC-200)	18 × 32 × 33 cm	30 × 40 × 50 cm
Weight	12.3 kg with DNA engine (PTC-200)	6.8 kg	~20 kg
Software	Opticon monitor 3.1	Opticon monitor 3.1	CFX manager Software, Security Edition. Comparison of up to 5000 samples. Can import files from all B/R real time PCR instruments
Additional features		Smallest Real Time PCR Instrument World Wide	Patent pending reduced mass block. Factory calibrated photonic shuttel (prewarmed). e-mail notification and data transfer via e-mail; USB Drive; Network enabeled
Heating/cooling method/accuracy/ uniformity			

Maufacturer	Applied Biosystems		
Model	"StepOne"	"StepOnePlus"	7300
Appearance	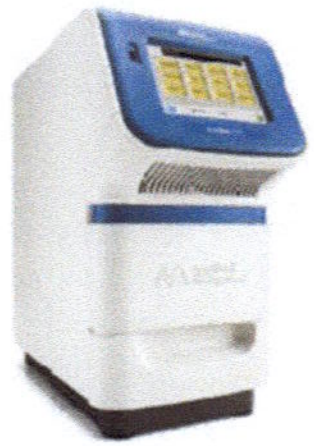	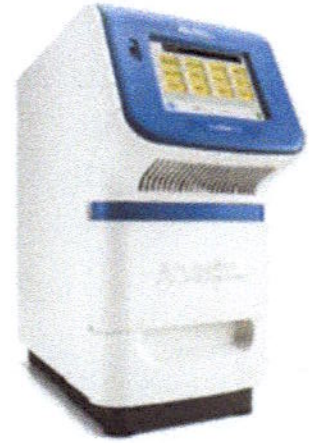	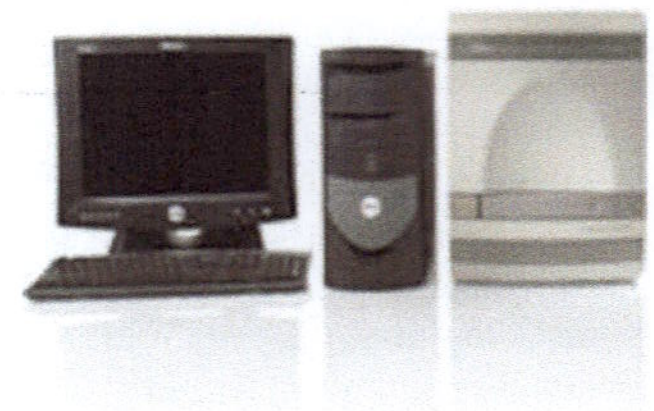
Detector	Photodiode	Photodiode	CCD
Light source	single exitation LED	single exitation LED	Lamp
Excitation channels	1	1	1
Emission (detection) channels	2 + 1 for ROX	3 + 1 for ROX	4
Targets	2 targets + ROX passive Reference dye	3 targets + ROX passive Reference dye	3 targets + ROX normalisation
Interchangeable filters	X	X	X
Multiple reference gene normalisation	X	X	X
Ramp rate	+/–1.6°C/s standard and 2.2°C/s FAST mode	+/–1.6°C/s standard and 2.2°C/s FAST mode	1.1°C/s heating and 1.1°C/s cooling
Adjustable ramp rate	√	Ö	X
Heating/cooling method	Peltier block	Peltier block	Peltier block
Gradient function	X	"VeriFlex block" 25°C (5°C zone-to-zone)	X
Sample format	48-well plate	96-well plate	96-well plate

Maufacturer	Applied Biosystems		
Model	"StepOne"	"StepOnePlus"	7300
Sample volume range	10–30 μL	10–30 mL	25–100 mL
Detection method	CCD camera	CCD camera	CCD camera
Data collection	48-well sequentiell LED array fire illuminate 1 well at a time	96-well sequentiell LED array fire illuminate 1 well at a time	Simultaneous
ROX reference dye needed	optional ROX passive reference dye (occupies 1 channel during run)	optional ROX passive reference dye (occupies 1 channel during run)	ROX passive reference dye (occupies 1 channel during run)
Open platform chemistry	ABI's TaqMan w. ROX chemistry	ABI's TaqMan w. ROX chemistry	ABI's TaqMan w. ROX chemistry
Dimensions (W × D × H)	25 × 51 × 43 cm	25 × 51 × 43 cm	34 × 45 × 49 cm
Weight	23.6 kg	24 kg	29 kg
Software	ABI	ABI	ABI
Additional features	LCD touchscreen, USB drive, remote monitoring and e-mail notification	LCD touchscreen, USB drive, remote monitoring and e-mail notification	
Heating/cooling method/accuracy/uniformity	Peltier block/+/–0,25°C/+/–0,5°C	Peltier block/+/–0,25°C/+/–0,5°C	Peltier block

Maufacturer	Applied Biosystems		Stratagene
Model	7500	7900HT	Stratagene Mx3000P MX3005P
Appearance	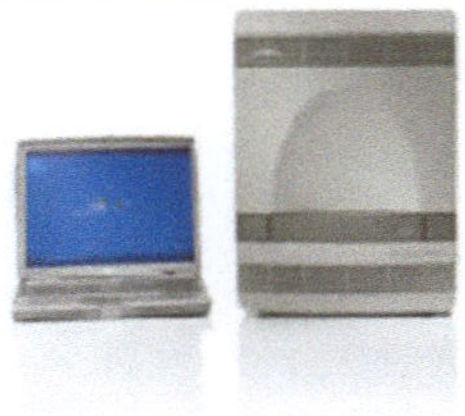	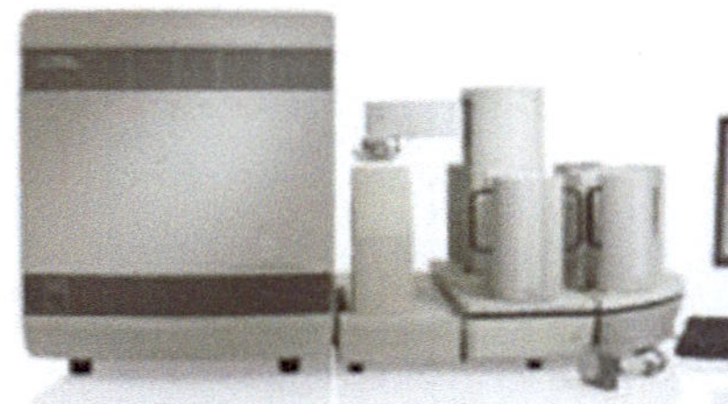	
Detector	CCD	CCD	CCD
Light source	Lamp	488 nm argon-ion laser (expensive and difficult to replace)	Quartz tungsten-halogen lamp
Excitation channels	5	1	4 or 5
Emission (detection) channels	5	4	4 or 5
Targets	4 targets + ROX normalisation	3 targets + ROX normalisation	Up to 4 or5
Interchangeable filters	X	X	Ö
Multiple reference gene normalisation	X	X	X
Ramp rate	Undetermined	Undetermined	Up to 2.5° C/s
Adjustable ramp rate	√	"Fast cycling" block upgrade now available: 96-well format only	X (dynamically controlled by the software)
Heating/cooling method	Peltier block	Peltier block	Peltier block
Gradient function	X	X	X

Maufacturer	Applied Biosystems		Stratagene
Model	7500	7900HT	Stratagene Mx3000P MX3005P
Sample format	96-well plate	96- and 384-well plate	96-well plate
Sample volume range	25–100 mL	5–50 mL; varies with block format	Optimised for 25 mL
Detection method	CCD camera	CCD camera	Photomultiplier tube (PMT)
Data collection	Simultaneous	Simultaneous	Scanning (takes up to 18 s)
ROX reference dye needed	ROX passive reference dye (occupies 1 channel during run)	ROX passive reference dye (occupies 1 channel during run)	Optional, results improve with ROX normalization
Open platform chemistry	ABI's TaqMan w. ROX chemistry	ABI's TaqMan w. ROX chemistry	Ö
Dimensions (W × D × H)	34 × 45 × 49 cm	125 × 84 × 64 cm	33 × 46 × 43 cm
Weight	34 kg	~80 kg	20 kg
Software	ABI	ABI	Stratagene
Additional features			
Heating/cooling method/accuracy/uniformity	Peltier block	Peltier block	Peltier block

Maufacturer	Roche		Corbett
Model	LightCycler 2.0	LightCycler 480	Rotor-Gene 3000
Appearance		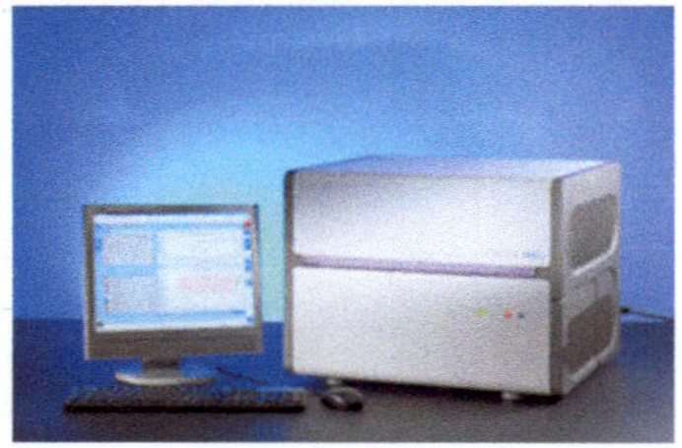	
Detector	PMT	CCD	PMT
Light source	470 nm blue LED (expensive and difficult to replace)	Xenon lamp	Carousel spins samples past detector (0.15 s)
Excitation channels	1	5	4
Emission (detection) channels	3	5	6
Targets	Up to 3	Up to 5	Up to 4
Interchangeable filters	X	X	X
Multiple reference gene normalisation	X	?	X
Ramp rate	Up to 20°C/s – but variable (depends on ambient temperature)	Fast ramping, but no detailed information	10°C/s – but variable (depends on ambient temperature)
Adjustable ramp rate	Variable but not controllable	?	Variable but not controllable

Maufacturer	Roche		Corbett
Model	LightCycler 2.0	LightCycler 480	Rotor-Gene 3000
Heating/cooling method	Air heated/cooled sample carousel	Peltier block	Air heated/cooled sample carousel
Gradient function	X (not possible with carousel)	X	(not possible with carousel)
Sample format	32 capillary carousel	96- and 384-well plate	36 (0.2 mL thin-walled tubes) or 72 (0.1 mL strip tubes)
Sample volume range	10–20 mL	20–100 μL (96 well) 5–20 μL (384 well)	Vary with rotor formats
Detection method	Photohybrid	CCD camera	Photomultiplier tube (PMT)
Data collection	Carousel spins all samples past detector (5 s)	Simultaneous	Carousel spins samples past detector (0.15 s)
ROX reference dye needed	X	X	X
Open platform chemistry	X (tied to LC chemistry)	X (tied to LC chemistry)	√
Dimensions (W × D × H)	45 × 30 × 40 cm	70 × 70 × 70 cm	38 × 48 × 31.5 cm
Weight	19.2 kg	59 kg	17 kg
Software	Roche LC soft	Roche LC soft	Corbett
Additional features			
Heating/cooling method/ accuracy/uniformity	Air heated/cooled sample carousel	Peltier block with special Liquid tech. for cooling	Air heated/cooled sample carousel

Maufacturer	Corbett	Eppendorf	
Model	Rotor-Gene 6000	Mastercycler ep realplex^2	Mastercycler ep
Appearance		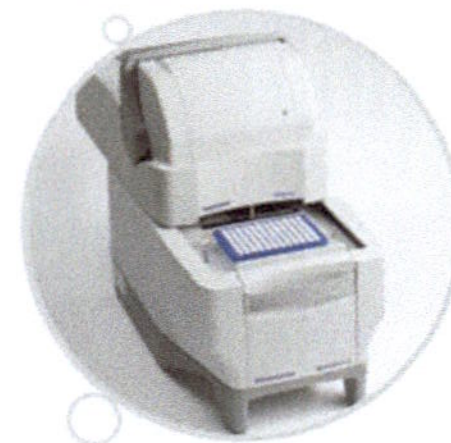	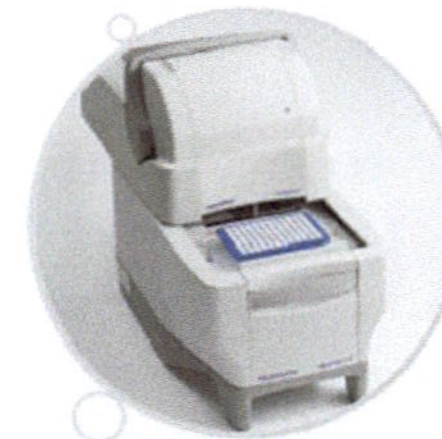
Detector	PMT	1 PMT	2 PMT
Light source	Carousel spins samples past detector (0.15 s)	470 nm blue LED array 96 optical fibers	470 nm blue LED array 96 optical fibers
Excitation channels	7	1	1
Emission (detection) channels	6	2	4
Targets	Up to 6	up to 2	up to 4?
Interchangeable filters	2, 5 or 6 Plex customized	X	X
Multiple reference gene normalisation	X	X	X
Ramp rate	10°C/s – but variable (depends on ambient temperature)	"CPM" photo-multipliers	"CPM" photo-multipliers

Maufacturer	Corbett	Eppendorf	
Model	Rotor-Gene 6000	Mastercycler ep realplex^2	Mastercycler ep
Adjustable ramp rate	Variable but not controllable	√	√
Heating/cooling method	Air heated/cooled sample carousel	Peltier block	Peltier block
Gradient function	(not possible with carousel)	√	√
Sample format	36 (0.2 mL thin-walled tubes) or 72 (0.1 mL strip tubes) or 100 (0.1 mL carussel)	96-well plate or 96 × 0.2mL tube	96-well plate or 96 × 0.2 mL tube
Sample volume range	Vary with rotor formats	25–100 mL	25–100 mL
Detection method	Photomultiplier tube (PMT)	Photomultiplier (CPM)	Photomultiplier (CPM)
Data collection	Carousel spins samples past detector (0.15 s)	96-well LED array illuminated 1 well at a time (? s for 96-well plate read)	96-well LED array illuminated 1 well at a time (? s for 96-well plate read)
ROX reference dye needed	X	X	X
Open platform chemistry	√	√	√
Dimensions (W × D × H)	38 × 48 × 31.5 cm	26 × 41 × 39.6 cm	26 × 41 × 39.6 cm
Weight	20 kg	24 kg	24 kg
Software	Corbett	Eppendorf	Eppendorf
Additional features			
Heating/cooling method/ accuracy/uniformity	Air heated/cooled sample carousel	Peltier block	Peltier block

Notice regarding Bio-Rad thermal cyclers and real-time systems: Chapter 8 is intended as a guideline to demonstrate similarities and differences. It is not a complete or exhaustive guide and the authors do not intend preference.

Purchase of this instrument conveys a limited non-transferable immunity from suit for the purchaser's own internal research and development and for use in applied fields other than Human In Vitro Diagnostics under one or more of U.S. Patents Nos. 5,656,493, 5,333,675, 5,475,610 (claims 1, 44, 158, 160–163 and 167 only), and 6,703,236 (claims 1–7 only), or corresponding claims in their non-U.S. counterparts, owned by Applera Corporation. No right is conveyed expressly, by implication or by estoppel under any other patent claim, such as claims to apparatus, reagents, kits, or methods such as 5' nuclease methods. Further information on purchasing licenses may be obtained by contacting the Director of Licensing, Applied Biosystems, 850 Lincoln Centre Drive, Foster City, California 94404, USA.

Bio-Rad's real-time thermal cyclers are licensed real-time thermal cyclers under Applera's United States Patent No. 6,814,934 B1 for use in research and for all other fields except the fields of human diagnostics and veterinary diagnostics.

Glossary of Terms

A

Absolute quantification The absolute quantification assay is used to quantitate unknown samples by interpolating their quantity from a standard curve

Allelic discrimination assay Assays designed to type for gene variants

AFLP Amplified fragment length polymorphism

Agarose gel electrophoresis A method for separating nucleic acids (DNA or RNA) within a gel made of agarose in a suitable buffer under the influence of an electrical field. Suitable for separation of large fragments of nucleic acid, separation is based primarily upon the size of the nucleic acid

Allele One of several alternate forms of a gene occupying a given locus on a chromosome or plasmid

Amino acids The 20 basic building blocks of proteins, consisting of the basic formula NH_2-CHR-COOH, where "R" is the side chain that defines the amino acid: Nonpolar side chains (hydrophobic), G Gly Glycine, A Ala Alanine, V Val Valine, I Ile Isoleucine, L Leu Leucine, F Phe Phenylalanine, P Pro Proline, M Met Methionine, W Trp Tryptophan, C Cys Cysteine, Noncharged polar side chains (hydrophilic), S Ser Serine, T Thr Threonine, Y Tyr Tyrosine, N Asn Asparagine, Q Gln Glutamine, Acidic side chains (very polar, hydrophilic), D Asp Aspartic Acid, E Glu Glutamic Acid, Basic side chains (very polar, hydrophilic), K Lys Lysine, R Arg Arginine, H His Histidine

Amino terminus Refers to the NH2 end of a peptide chain (by custom drawn at the left of a protein sequence)

Amplicon The product' of double-stranded DNA created during PCR

Amplification The production of additional copies of a chromosomal sequence, found either as intrachromosomal or extrachromosomal DNA. Also refers to the in vitro process in the polymerase chain reaction

E.A. Pestana et al., *Early, Rapid and Sensitive Veterinary Molecular Diagnostics - Real Time PCR Applications*, DOI 10.1007/978-90-481-3132-7,

Amplification plot The plot of Fluorescence signal versus cycle numberwhich correlates with the initial amount of target nucleic acid during the exponential phase of PCR

Amplimer Region of DNA sequence that is amplified during a PCR reaction and that is defined by a pair of PCR primers (these primer pairs are sometimes called amplimers)

AMV Avian myeloblastosis virus RT

Anchor sequence A hydrophobic amino acid sequence that fixes a segment of a newly synthesized, translocating protein within the lipid bilayer membrane of the endoplasmic reticulum

Annealing The binding of complementary DNA or RNA sequences between bases via hydrogen bonding

Annealing temperature The temperature at which a length of single-strand (heat denatured) DNA or RNA will anneal to a complementary strand. Lower temperatures may permit non-specific binding.

Antisense strand (or primer) Refers to the RNA or DNA strand of a duplex molecule that is complementary to that encoding a polypeptide. More specifically, the DNA strand that serves as template for the synthesis of RNA and which is complementary to it. "Antisense oligonucleotides" hybridize to mRNA, and are used to prime cDNA synthesis

Avidin A glycoprotein that binds to biotin with very high affinity (Kd = 10–15)

B

Back mutation Reverse the effect of a point or frame-shift mutation that had altered a gene; thus it restores the wild-type phenotype (see REVERTANT)

Bacteriophage A virus that infects bacteria; often simply called a phage. The phages that are most often used in molecular biology are the E. coli viruses lambda, M13 and T7

Base Adenine (A), cytosine (C), guanine (G) and thymine (T) are the molecules composing DNA. These are bases in DNA that form two pairs of complementary molecules, hydrogen bonds can link adenine to thymine (A_T) and guanine to cytosine (G_C). If the sequence of bases matches a complementary sequence in a second strand of DNA the bonds can hold the two strands together; thus forming double-stranded

Base pair (bp) One pair of complementary nucleotides within a duplex strand of a nucleic acid (BP) and 103 bp = kb) Under Watson-Crick rules, these pairs consist of one pyrimidine and one purine: i.e., C-G, A-T (DNA) or A-U (RNA). However, "noncanonical" base pairs (e.g., G-U) are common in RNA secondary structure

Baseline The initial cycles of PCR during which there is little change in fluorescence signal (usually cycles 3–15)

Biotin A coenzyme that is essential for carboxylation reactions (see Avidin)

Blunt end A terminus of a duplex DNA molecule that ends precisely at a base pair, with no overhang (unpaired nucleotide) in either strand. Some but not all restriction endonucleases leave blunt ends after cleaving DNA. Blunt-ended DNA can be ligated nonspecifically to other blunt-ended DNA molecules (compare with STICKY END)

Box Refers to a short nucleic acid consensus sequence or motif that is universal within kingdoms of organisms. Examples of DNA boxes are the Pribow box (TATAAT) for RNA polymerase, the Hogness box (TATA) that has a similar function in eukaryotic organisms, and the homeo box. RNA boxes have also been described, such as Pilipenko's Box-A motif that may be involved in ribosome binding in some viral RNAs

BSA Bovine serum albumin

C

CAP A 7-methyl guanosine residue linked 5′ to 5′ through a triphosphate bridge to the 5′ end of eukaryotic mRNAs; facilitates initiation of translation

Carboxyl terminus Refers to the COOH end of a peptide chain (by custom drawn at the right of a protein sequence)

cDNA Complementary DNA. A DNA molecule that was originally copied from an RNA molecule by reverse transcription. The term "cDNA" is commonly used to describe double-stranded DNA that originated from a single-stranded RNA molecule, even though only one strand of the DNA is truly complementary to the RNA

cDNA Library A collection of cDNA fragments, each of which has been cloned into a separate vector molecule

Chaperone proteins A series of proteins present in the endoplasmic reticulum that guide the proper folding of secreted proteins through a complex series of binding and release reactions

Chromosome walking The sequential isolation of clones carrying overlapping sequences of DNA that span large regions of a chromosome. Overlapping regions of clones can be identified by hybridisation

Clone Describes a large number of cells, viruses, or molecules that are identical and derived from a single ancestral cell, virus or molecule. The term can be used to refer to the process of isolating single cells or viruses and letting them proliferate (as in a hybridoma clone, which is a "biological clone"), or the process of isolating and replicating a piece of DNA by recombinant DNA techniques ("molecular clone"). The use of the word as a verb is acceptable for the former meaning, but not necessarily the latter meaning

CIS As used in molecular biology, an interaction between two sites that are located within the same molecule. However, a cis-acting protein can either be one that acts only on the molecule of DNA from which it was expressed, or a protein that acts on itself (e.g., self-proteolysis)

Cistron A nucleic acid segment corresponding to a polypeptide chain, including the relevant translational start (initiation) and stop (termination) codons

Codon A nucleotide triplet (sequence of three nucleotides) that specifies a specific amino acid, or a translational start or stop

Codon bias The tendency for an organism or virus to use certain codons more than others to encode a particular amino acid. An important determinant of codon bias is the guanosine-cytosine (GC) content of the genome. An organism that has a relatively low G+C content of 30% will be less likely to have a G or C at the third position of a codon (wobble position) than a A or T to specify an amino acid that can be represented by more than one codon

Competent Bacterial cells that are capable of accepting foreign extra-chromosomal DNA. There are a variety of processes by which cells may be made competent

Coefficient of variation (CV) A statistical representation of the precision of a test

Ct (thershold cycle) The fractional cycle number at which the fluorescence passes the fixed threshold. It is inversely correlated to the logarithm of the initial copy number. The Ct value assigned to a particular well thus reflects the point during the reaction at which a sufficient number of amplicons have accumulated

CFT Complement Fixation Test. Serological technique whereby a fixed system of sheep red blood cells, antibody to sheep red blood cells and complement (from guinea pig sera) is titrated using a heamolysis assay. The system is challenged by the addition of antigen/antibody reactions to the titrated complement. Any reaction uses (fixes) the complement and this is assessed in the pre-titrated system where a lack of heamolysis indicates extent of test antigen/antibody reaction

Consensus sequence A linear series of nucleotides, commonly with gaps and some degeneracy that define common features of homologous sequences or recognition sites for proteins that act on or bind to nucleic acids

Conservative substitution Nucleotide mutation that alters the amino acid sequence of the protein, but which causes the substitution of one amino acid with another that has a side chain with similar charge/polarity characteristics (see AMINO ACID). The size of the side chain may also be an important consideration. Conservative mutations are generally considered unlikely to profoundly alter the structure or function of a protein, but there are many exceptions

Conserved Similar in structure or function

Contamination Extraneous DNA present in the sample or reaction that will be amplified and give a false positive result

Contig A series of two or more individual DNA sequence determinations that overlap. In a sequencing project the contigs get larger and larger until the gaps between the contigs fill

Cosmid A genetically engineered plasmid containing bacteriophage lambda packaging signals and potentially very large pieces of inserted foreign DNA (up to 50 kb) that can be replicated in bacterial cells. Cosmid cloning allows for isolation of DNA fragments that are larger than those that can be cloned in conventional plasmids

D

Database search Once an open reading frame or a partial amino acid sequence has been determined, the investigator compares the sequence with others in the databases using a computer and a search algorithm. This is usually done in a protein database such as PIR or Swiss-Prot. Nucleic acid sequences are in GenBank and EMBL databases. The search algorithms most commonly used are BLAST and FASTA

DC Direct current

Degeneracy Refers to the fact that multiple different codons in mRNA can specify the same amino acid in an encoded protein

Delta Rn (ΔRn) The magnitude of the fluorescence signal generated during the PCR at each time point

Denaturation With respect to nucleic acids, refers to the conversion from double-stranded to the single-stranded state, often achieved by heating or alkaline conditions. This is also called "melting" DNA. With respect to proteins, refers to the disruption of tertiary and secondary structure, often achieved by heat, detergents, chaotropes, and sulfhydryl-reducing agents

Denaturing gel An agarose or acrylamide gel run under conditions that destroy secondary or tertiary protein or RNA structure. For protein, this usually means the inclusion of 2-ME (that reduces disulfide bonds between cysteine residues) and SDS and/or urea in an acrylamide gel. For RNA, this usually means the inclusion of formaldehyde or glyoxal to destroy higher ordered RNA structures. In DNA sequencing gels, urea is included to denature dsDNA to ssDNA strands. In denaturing gels, macromolecules tend to be separated on the basis of size and (to some extent) charge, while shape and oligomerization of molecules are not important. Contrast with NATIVE GEL

Deoxyribonuclease (DNase) An enzyme that specifically catalyzes the hydrolysis of DNA

Deoxyribonucleotide Nucleotides that are the building blocks of DNA and that lack the 2′ hydroxyl moiety present in the ribonucleotides of RNA

DEPC Diethyl pyrocarbonate

Dideoxyribonucleotide (ddntp) A nucleotide (dATP, dTP, dCTP and dTTP) that lacks both 3′ and 2′ hydroxyl groups. Such dideoxynucleotides can be added to a growing nucleic acid chain, but do not then present a 3′ –OH group that can support further propagation of the nucleic acid chain. Thus such compounds are also called "chain terminators", and are useful in DNA and RNA (ATP, GTP, CTP and UTP) sequencing reactions (see DEOXYRIBO-NUCLEOTIDE)

Dideoxy Sequencing Enzymatic determination of DNA or RNA sequence by the method of Sanger and colleagues, based on the incorporation of chain terminating dideoxynucleotides in a growing nucleic acid strand copied by DNA polymerase or reverse transcriptase from a DNA or RNA template. Separate reactions include dideoxynucleotides containing A, C, G, or T bases. The reaction products represent a collection of new-labelled DNA strands of varying lengths, all terminating with a dideoxynucleotide at the 3′ end (at the site of a complementary base in the template nucleic acid), and are separated in a polyacrylamide/urea gel to generate a sequence "ladder". This method is more commonly used than "Maxam-Gilbert" (chemical) sequencing

Direct repeats Identical or related sequences present in two or more copies in the same orientation in the same molecule of DNA; they are not necessarily adjacent

DMSO Di methyl suphoxide

DNA ligase An enzyme (usually from the T4 bacteriophage) that catalyzes formation of a phosphodiester bond between two adjacent bases from double-stranded DNA fragments. RNA ligases also exist, but are rarely used in molecular biology

DNA polymersase An enzyme that synthesizes a double stranded DNA molecule using single strand DNA primer as template

Dot Blot DNA or RNA is simply spotted onto nitrocellulose or nylon membranes, denatured and hybridized with a probe. Unlike Southern or northern blots, there is no separation of the target DNA or RNA by electrophoresis (size), and thus potentially much less specificity

DOP PCR Degenerate Oligonucleotide Primer-PCR

Downstream Identifies sequences proceeding farther in the direction of expression; for example, the coding region is downstream from the initiation codon, toward the 3′ end of an mRNA molecule. Sometimes used to refer to a position within a protein sequence, in which case downstream is toward the carboxyl end that is synthesized after the amino end during translation

DsDNA-binding agent A molecule that emits fluorescence when bound to dsDNA

Duplex A nucleic acid molecule in which two strands are base paired with each other

Dynamic range The range of initial template concentrations over which accurate Ct values are obtained

E

EDTA Ethylinediaminetatraacetic acid

Electroporation A method for introducing foreign nucleic acid into bacterial or eukaryotic cells that uses a brief, high voltage DC charge that renders the cells permeable to the nucleic acid. Also useful for introducing synthetic peptides into ÿukaryotic cells

Efficiency of the reaction The efficiency of the reaction can be calculated by the following equation: $E = 10(-1/\text{slope}) -1$. When the efficiency of the PCR is 90–100%, doubling of the amplicon is expected at each cycle

ELISA Enzyme Linked Immunosorbent Assay. Very widely used group of methods whereby an enzyme attached to reactant in a system is used to generate a colour signal with an appropriate chromogen and substrate combination. Methods include Indirect, sandwich and competitive systems. Main serological tool for diagnosis. Most methods employ microplate methods utilizing antigens or antibodies passively attached to microplate wells. Separation of sequential reactions is easily made by washing steps. Multichannel spectrophotometers make reading colour rapid

Elongation or extension Phase of PCR cycle following annealing of primer during that the Taq polymerase manufactures a strand of DNA. The optimum temperature depends on the enzyme used but is usually between 68 and 72°C

ELOSA Enzyme linked oligonucleotide sorbent assay

End labeling The technique of adding a radioactively labeled group to one end ($5'$ or $3'$ end) of a DNA strand

Endonuclease Cleaves bonds within a nucleic acid chain; they may be specific for RNA or for single-stranded or double-stranded DNA. A restriction enzyme is a type of endonuclease

End-point analysis Analysis of the data collected at the end-point phase. Usually used in allelic discrimination assays (genotyping)

Enhancer A eukaryotic transcriptional control element that is a DNA sequence that acts at some distance to enhance the activity of a specific promoter sequence. Unlike promoter sequences, the position and orientation of the enhancer sequence is generally not important to its activity

Endogenous control This is an RNA or DNA that is naturally present in each experimental sample

Exogenous control This is a characterized RNA or DNA spiked into each sample at a known concentration

Ethidium bromide Intercalates within the structure of nucleic acids in such a way that they fluoresce under UV light. Ethidium bromide staining is commonly used to visualize RNA or DNA in agarose gels placed on UV light boxes. Proper precautions are required, because the ethidium bromide is highly mutagenic and the UV light damaging to the eyes. Ethidium bromide is also included in cesium chloride gradients during ultracentrifugation, to separate supercoiled circular DNA from linear and relaxed circular DNA

Evolutionary clock Defined by the rate at which mutations accumulate within a given gene

Exon The portion of a gene that is actually translated into protein

Exonuclease An enzyme that hydroylses DNA beginning at one end of a strand, releasing nucleotides one at a time (thus, there are 3′ or 5′ exonucleases)

Expression Usually used to refer to the entire process of producing a protein from a gene, that includes transcription, translation, post-translational modification and possibly transport reactions

Expression vector A plasmid or phage designed for production of a polypeptide from inserted foreign DNA under specific controls. Often an inducer is used. The vector always provides a promoter and often the transcriptional start site, ribosomal binding sequence, and initiation codon. In some cases the product is a fusion protein

F

FAM 6-carboxy fluorescein. Most commonly used reporter dye at the 5′ end of a TaqMan® probe

Fluorescence resonance energy transfer (FRET) The interaction between the electronic excited states of two dye molecules. The excitation is transferred from one (the donor) dye molecule to the other (the acceptor) dye molecule. FRET is distance-dependent and occurs when the donor and the acceptor dye are in close proximity

Footprinting A technique for identifying the site on a DNA (or RNA) molecule that is bound by some protein by virtue of the protection afforded phosphodiester bonds in this region against attack by nuclease or nucleolytic compounds

Frameshift mutation A mutation (deletion or insertion, never a simple substitution) of one or more nucleotides but never a multiple of 3 nucleotides, that shortens or lengthens a trinucleotide sequence representing a codon; the result is a shift from one reading frame to another reading frame. The amino acid sequence of the protein downstream of the mutation is completely altered, and may even be much shorter or longer due to a change in the location of the first termination (stop) codon

Fusion protein A product of recombinant DNA in which the foreign gene product is juxtaposed ("fused") to either the carboxyl-terminal or amino-terminal portion of a polypeptide encoded by the vector itself. Use of fusion proteins often facilitates expression of otherwise lethal products and the purification of recombinant proteins

G

gDNA Genomic DNA. Double helix containing the genetic blueprint of the organism. DNA sugar phosphate "backbone" molecules are joined by the pairing of bases, thymine with adenine and cytosine with guanine. Hydrogen bonds form between the complementary bases holding the two strands together

Gel electrophoresis Amplicons of differing lengths can be separated by their migration through an agarose gel whilst subject to an electrical field. The negatively charged DNA molecules move at a rate that is inversely proportional to their size. The DNA can be ÿybridizat by staining with a fluorescent dye and viewing under UV light. The size of the amplicons is predicted by the positioning of the primers on the target DNA and a product or "band" of the correct size is taken as positive result. Confirmation can be provided by the use of specific probes to check the identity of the band

Gel shift A method by which the interaction of a nucleic acid (DNA or RNA) with a protein is detected. The mobility of the nucleic acid is monitored in an agarose gel in the presence and absence of the protein: if the protein binds to the nucleic acid, the complex migrates more slowly in the gel (hence "gel shift"). A "supershift" allows determination of the specific protein, by virtue of a second shift in mobility that accompanies binding of a specific antibody to the nucleic acid-protein complex

Gene Generally speaking, the genomic nucleotide sequence that codes for a particular polypeptide chain, including relevant transcriptional control sequences and introns (if a eukaryote). However, the term is often loosely used to refer to only the relevant coding sequence

Gene conversion The alteration of all or part of a gene by a homologous donor DNA that is itself not altered in the process

Genome The complete set of genetic information defining a particular animal, plant, organism or virus

Genomic library A DNA library that contains DNA fragments hopefully representing each region of the genome of an organism, virus, etc, cloned into individual vector molecules for subsequent selection and amplification. The DNA fragments are usually very small in size compared with the genome. Such libraries are cDNA libraries when prepared from RNA viruses

Genotype The genetic constitution of an organism; determined by its nucleic acid sequence. As applied to viruses, the term implies a group of evolutionarily related viruses possessing a defined degree of nucleotide sequence relatedness

GLP Good laboratory practice

Glycoprotein A glycosylated protein

Glycosylation The covalent addition of sugar ÿybridiz to N or O atoms present in the side chains of certain amino acids of certain proteins, generally occurring within the Golgi apparatus during secretion of a protein. Glycosylation sites are only partially predictable by current computer searches for relevant motifs in protein sequence. Glycosylation may have profound but very unpredictable effects on the folding, stability, and antigenicity of secreted proteins. Glycosylation is a property of eukaryotic cells, and differs among different cell types (i.e., it may be very different in yeast or insect cells used for protein expression, when compared with Chinese hamster ovary (CHO) cells)

H

HA Haemagglutination

Hairpin A helical (duplex) region formed by base pairing between adjacent (inverted) complementary sequences within a single strand of RNA or DNA

House keeping genes Genes that are widely expressed in abundance and are usually used as reference genes for normalization in real-time PCR with the assumption of "constant expression"

Heteroduplex DNA Generated by base pairing between complementary single strands derived from different parental duplex molecules; heteroduplex DNA molecules occur during genetic recombination in vivo and during hydridization of different but related DNA strands in vitro. Since the sequences of the two strands in a heteroduplex differ, the molecule is not perfectly base-paired; the melting temperature of a heteroduplex DNA is dependent upon the number of mismatched base pairs

Homologous recombination The exchange of sequence between two related but different DNA (or RNA) molecules, with the result that a new "chimeric" molecule is created. Several mechanisms may result in recombination, but an essential requirement is the existence of a region of homology in the recombination partners. In DNA recombination, breakage of single strands of DNA in the two recombination partners is followed by joining of strands present in opposing molecules, and may involve specific enzymes. Recombination of RNA molecules may occur by other mechanisms

Homology Indicates similarity between two different nucleotide or amino acid sequences, often with potential evolutionary significance. It is probably better to use more quantitative and descriptive terms such as nucleotide "identity" or, in the case of proteins, amino acid "identity" or "relatedness" (the latter refers to the presence of amino acids residues with similar polarity/charge characteristics at the same position within a protein)

Hybridisation The process of base pairing leading to formation of duplex RNA or DNA or RNA-DNA molecules

Hybridisation probes One of the main fluorescence-monitoring systems for DNA amplification. Melting curve analysis is possible with hybridization probes

Hydrolysis probes Probes that are hydrolyzed by the 5′ endonuclease activity of Taq Polymerase during PCR

Hot start To ÿybridiz mismatched primers interfering with the PCR the reagents are heated to above the annealing temperature before all the reaction components are allowed to mix. This is often achieved by using wax to separate the primers from the enzyme, the reaction components can mix only when the wax melts thus avoiding the elongation of non-specifically bound primers. Alternatively, enzymes are now available that will only elongate at elevated temperatures

Hydrophilicity plot A computer plot that examines the relative summed hyrophobicity/hydrophilicity of adjacent amino acid sidechains (usually within a moving window of about 6 amino acid residues) along the primary sequence of a polypeptide chain. Values for the contribution of sidechains of each the 20 common amino acids to hydrophobicity/hydrophilicity have been developed by Hopp & Woods, and Kyte & Doolittle, and these plots are often named after these workers. Generally, hydrophobic regions of proteins are considered likely to be in the interior of the native protein, while hydrophilic domains are likely to be exposed on the surface and thus possibly antigenic sites (epitopes). At best, these are crude predictions

I

Internal positive control (IPC) Exogenous sample added to a multiplex assay or run on its own to monitor the presence of inhibitors in the template

Immunoblot See Western blot

Inducer A small molecule, such as IPTG, that triggers gene transcription by binding to a regulator protein, such as LacZ

Inhibition The exponential nature of PCR means that a small change in the efficiency of each cycle can result in a large difference to the final product. Inhibition may result from chemical interference in the action of polymerase or other components of the reaction and could be removed by further purification of the sample. Alternatively, a large excess of DNA in the sample may prevent efficient amplification of the target and further dilution of sample is required. Checks for inhibition can be made by adding a known target to the sample to demonstrate the efficiency of the amplification

Initiation codon The codon at which translation of a polypeptide chain is initiated. This is usually the first AUG triplet in the mRNA molecule from the 5′ end, where the ribosome binds to the cap and begins to scan in a 3′ direction. However, the surrounding sequence context is important and may lead to the first

AUG being bypassed by the scanning ribosome in favor of an alternative, downstream AUG. Also called a "start codon". Occasionally other codons may serve as initiation codons, e.g. UUG

Insert Foreign DNA placed within a vector molecule

Insertion sequence A small bacterial transposon carrying only the genetic functions involved in transposition. There are usually inverted repeats at the ends of the insertion sequence

Intron Intervening sequences in eukaryotic genes that do not encode protein but which are transcribed into RNA. Removed from pre-mRNA during nuclear splicing reactions

Inverted repeats Two copies of the same or related sequence of DNA repeated in opposite orientation on the same molecule (contrast with DIRECT REPEATS). Adjacent inverted repeats constitute a palindrome

In situ PCR PCR of DNA within histological specimens

IRE PCR Interspersed repetitive element (IRE)-PCR

ISO A network of national standards institutes from 148 countries working in partnership with international organizations, governments, industry, business and consumer representatives

Isothermal amplification Amplification step proceeds at one temperature and does not require thermocycling apparatus

K

Kilobase Unit of 1000 nucleotide bases, either RNA or DNA

Klenow fragment The large fragment of E. coli DNA polymerase I that lacks $5'$ -> $3'$ exonuclease activity. Very useful for sequencing reactions, that proceed in a $5' > 3'$ fashion (addition of nucleotides to templated free $3'$ ends of primers)

Knock-out The excision or inactivation of a gene within an intact organism or even animal (e.g., "knock-out mice"), usually carried out by a method involving homologous recombination

L

LCR Ligase chain reaction. Amplification of DNA by repeated "cycling" reactions using ligase enzyme to "ligate" or "join" pairs of primers. Covered by patents owned by Abbott Laboratories (USA). A range of diagnostic kits is available

Library A set of cloned fragments together representing with some degree of redundancy the entire genetic complement of an organism (see cDNA LIBRARY, GENOMIC LIBRARY)

Ligase See DNA ligase

Ligation See DNA ligase

Linear view Amplification plot view displayed using exact ΔRn values on the Y-axis. Used for threshold setting

Linkage The tendency of genes to be inherited together as a result of their relatively close proximity on the same chromosome, or location on the same plasmid

Linker A short oligodeoxyribonucleotide, usually representing a specific restriction endonuclease recognition sequence, that may be ligated onto the termini of a DNA molecule to facilitate cloning. Following the ligation reaction, the product is digested with the endonuclease, generating a DNA fragment with the desired sticky or blunt ends

Lipofectin A commercially marketed liposome suspension that is mixed with DNA or RNA to facilitate uptake of the nucleic acid by eukaryotic cells (see TRANSFECTION)

Long PCR PCR in which the product is several kilobases long. Using cocktails of enzymes products of up to 40 kb have been accurately manufactured

LSD

M

mRNA Messenger RNA

Melting The dissociation of a duplex nucleic acid molecule into single strands, usually by increasing temperature. See DENATURATION

Melting curve analysis When DNA-binding dyes are used, as the fragment is heated, a sudden decrease in fluorescence is detected when Tm is reached (due to dissociation of DNA strands and release of the dye). This point is determined from the inflection point of the melting curve or the melting peak of the derivative plot

$MgCl_2$ Magnesium chloride

Mg 2+ Magnesium ion

Missense mutation A nucleotide mutation that results in a change in the amino acid sequence of the encoded protein (contrast with SILENT MUTATION)

Minor groove binders (MGBs) These dsDNA-binding agents are attached to the 3′ end of TaqMan® probes to increase the Tm value (by stabilization of hybridization) and to design shorter probes. Longer probes reduce design flexibility and are less sensitive to mismatch discrimination. MGBs also reduce background fluorescence and increase dynamic range due to increased efficiency of reporter quenching (these probes use non-fluorescent quenchers at the 3′ end). By allowing the use of shorter probes with higher Tm values, MGBs enhances mismatch discrimination in genotyping assays

M-MuLV < Moloney murine leukemia virus RT

Molecular beacons Hairpin probes consisting on a sequence-specific loop region flanked by two inverted repeats. Reporter and quencher dyes are attached to each end of the molecule and remain in close contact unless sequence-specific binding occurs and reporter emission (FRET) occurs

Motif A recurring pattern of short sequence of DNA, RNA, or protein, that usually serves as a recognition site or active site. The same motif can be found in a variety of types of organisms mRNA–A cytoplasmic RNA that serves directly as the source of code for protein synthesis

Multicistronic message An mRNA transcript with more than one cistron and thus encoding more than one polypeptide. These generally do not occur in eukaryotic organisms, due to differences in the mechanism of translation initiation

Multicopy plasmids Present in bacteria at amounts greater than one per chromosome. Vectors for cloning DNA are usually multicopy; there are sometimes advantages in using a single copy plasmid

Multiple cloning site An artificially constructed region within a vector molecule that contains a number of closely spaced recognition sequences for restriction endonucleases. This serves as a convenient site into which foreign DNA may be inserted

Multiplex PCR PCR with two or more independent sets of primers in the same reaction tube

N

N terminus See AMINO TERMINUS

Native gel An electrophoresis gel run under conditions that do not denature proteins (i.e., in the absence of SDS, urea, 2-mercaptoethanol, etc.)

Nested PCR (n-PCR) A very sensitive method for ÿybridization of DNA, that takes part of the product of a single PCR reaction (after 30–35 cycles), and subjects it to a new round of PCR using a different set of PCR primers which are nested within the region flanked by the original primer pair (see Polymerase Chain Reaction)

Nick In duplex DNA, this refers to the absence of a phosphodiester bond between two adjacent nucleotides on one strand

Nick translation A method for introducing labeled nucleotides into a double-stranded DNA molecule that involves making small nicks in one strand with Dnase, and then repairing with DNA polymerase I

Nonconservative Substitution A mutation that results in the substitution of one amino acid within a polypeptide chain with an amino acid belonging to a different polarity/charge group

Nonsence codon See stop codon

Nonsense mutation A change in the sequence of a nucleic acid that causes a nonsense (stop or termination) codon to replace a codon representing an amino acid

Nontranslated RNA (NTR) The segments located at the 5′ and 3′ ends of a mRNA molecule that do not encode any part of the polyprotein; may contain important translational control elements

Normalisation The use of a control gene that is expressed at a constant level to normalize the gene expression results for variable template amount or template quality

Northern blot RNA molecules are separated by electrophoresis (usually in an agarose gel) on the basis of size, then transferred to a solid-phase support (nitrocellulose paper or suitable other membrane) and detected by ÿybridization with a labeled probe (see SOUTHERN BLOT, WESTERN BLOT)

No template control (NTC) NTC includes all of the RT-PCR reagents except the RNA template. No product should be synthesized in the NTC or NAC; if a product is amplified, this indicates contamination (fluorescent or PCR products) or presence of genomic DNA in the RNA sample. NTC is not equivalent to H_2O controls and H_2O controls are not used in qPCR experiments

Nucleoside The composite sugar and purine or pyrimidine base that is present in nucleotides which are the basic building blocks of DNA and RNA. Compare with NUCLEOTIDE: Nucleoside = Base + Sugar

Nucleotide The composite phosphate, sugar, and purine or pyrimidine base that are the basic building blocks of the nucleic acids DNA and RNA. The five nucleotides are adenylic acid, guanylic acid (contain purine bases), and cytidylic acid, thymidylic acid, and uridylic acid (contain pyrimidine bases)

NASBA Nucleic Acid Sequence-Based Amplification. Amplification of RNA using RNA polymerase. Developed by Cangene Corporation (Canada), also licensed to Organon Teknika (Belgium). Nucleotide = Base + Sugar + Phosphate (1, 2, or 3)

O

OIE Office International des Epizootie, World Organisation for Animal health. 12, rue de Prony, 75017 Paris, France. Email : oie@o ie.int; Web Site: www.oie.int

Oligodeoxyribonucleotide A short, single-stranded DNA molecule, generally 15–50 *nucleotides* in length, that may be used as a *primer* ÿybridization*sation probe*. Oligodeoxyribonucleotides are synthesized chemically under automated conditions

Oligonucleotide See OLIGODEOXYRIBONUCLEOTIDE

Oncogene One of a number of genes believed to be associated with the malignant transformation of cells; originally identified in certain oncogenic retroviruses (*v-onc*) but also present in cells (*c-onc*)

One-tube nested PCR Both reactions in a nested PCR are performed concurrently in one tube without the need to add further reagents

Open reading frame A region within a reading frame of an mRN" molecule that potentially encodes a *polypeptide*; and that does not contain a *translational stop codon* (see READING FRAME)

Operator The site on DNA at which a *repressor* protein binds to prevent *transcription* from initiating at the adjacent *promoter*

Operon A complete unit of bacterial gene expression and regulation, including the structural gene or genes, regulator gene(s), and control elements in DNA recognized by regulator gene products(s)

Origin A site within a DNA sequence of a chromosome, plasmid, or non-integrated virus at which replication of the DNA is initiated

Overhang A terminus of a *duplex* DNA molecule that has one or more unpaired nucleotides in one of the two strands (hence eith′r a 3′ or 5′ overhang). Cleavage of DNA with many restriction endonucleases leaves such overhangs (see STICKY END)

P

Package In recombinant DNA procedures, refers to the step of incorporation of *cosmid* or other lambda *vector* DNA with an *insert* into a *phage* head for transduction of DNA into host

Page Polyacrylamide gel electrophoresis

Palindromic sequence A nucleotide sequence that is the same when read in either direction, usually consisting of adjacent inverted repeats. *Restriction endonuclease recognition sites* are palindromes:

Passive reference dye A fluorescence dye that provides an internal reference to which the reporter dye signal can be normalized during data analysis by the software, to correct for fluctuations from well to well caused by changes in concentration or volume

PCR See Polymerase Chain Reaction

PCR-ELIS PCR Enzyme linked immunosorbent assay

Pfu Plaque forming units

Phenotype The appearance of other characteristics of an organism resulting from the interaction of its genetic constitution with the environment

Phosphatase alkaline An enzyme that catalyzes the hydrolysis of phosphomonoesters of the 5′ nucleotides. Used to dephosphorylate (remove phosphate groups from) the 5′ ends of DNA or RNA molecules, to facili′ate 5′ end-labeling with 32P added back by T4 polynucleotide kinase; or to dephosphorylate the 5′ ends of DNA molecules to prevent unwanted *ligation* reactions during cloning

Phosphodiester bond The covalent bond between the 3′ hydroxyl in the sugar ring of one nucleotide and the 5′ phosphate group of the sugar ring of the adjacent nucleotide residue within a nucleic acid: 5′-Ribose- 3′ – O – P(O)2 – O – 5′ – Ribose – 3′ etc

Phosphorylation The addition of a phosphate monoester to a macromolecule, catalyzed by a specific kinase enzyme. With respect to proteins, certain *amino acid* side chains (serine, threonine, tyrosine) are subject to phosphorylation catalyzed by protein kinases; altering the phosphorylation status of a protein may have dramatic effects on its biologic properties, and is a common cellular control mechanism. With respect to DNA, 5′ ends must be phosphorylated for *ligation*

Plasmid An extrachromosomal, usually circular, double-stranded DNA that is capable of *replication* within a cell, and which usually contains and expresses genes encoding resistance to antibiotics. By strict definition, a plasmid is not essential to the life of the cell

Point mutation A single nucleotide substitution within a gene; there may be several point mutations within a single gene. Point mutations do not lead to a shift in reading frames, thus at most cause only a single amino acid substitution (see FRAMESHIFT MUTATION)

Polyacrylamide gel (PAGE) Used to separate proteins and smaller DNA fragments and *oligonucleotides* by electrophoresis. When run under conditions that denature proteins (i.e., in the presence of 2-mercaptoethanol, SDS, and possibly urea), molecules are separated primarily on the basis of size

Poly-A track A lengthy adenylic acid polymer (RNA) that is covalently linked to the 3′ end of newly synthesized *mRNA* molecules in the nucleus

Polymerase chain reaction (PCR) A DNA amplification reaction involving multiple (30 or more) cycles of *primer annealing, extension*, and denaturation, usually using a heat-stable DNA *polymerase* such as *Taq polymerase*. Paired primers are used, that are complementary to opposing strands of the DNA and which flank the area to be amplified. Under optimal conditions, single DNA sequence can be amplified a million-fold. The basic PCR process is covered by patents owned by Hoffman-La Roche Inc and Hoffman-La Roche Ltd. A range of diagnostic kits is available

Polymerase An enzyme that catalyzes the addition of a *nucleotide* to a nucleic acid molecule. There is a wide variety of RNA and DNA polymerases that have a wide range of specific activities and that operate optimally under different conditions. In general, all polymerases require *templates* upon which to build a new strand of DNA or RNA; however, DNA polymerases also require a *primer* to initiate the new strand, while RNA polymerases start synthesis at a specific *promoter* sequence

Polymorphism Variation within a DNA or RNA sequence

Polynucleotide kinase Enzyme which catalyzes the transfer of the terminal phosphate of ATP to 5′ hydroxyl termini of polynucleotides, either DNA or RNA. Usually derived from T4 bacteriophage. pre-mRNA – An RNA molecule that is *transcribed* from chromosomal DNA in the nucleus of eukaryotic cells, and subsequently processed through *splicing* reactions to generate the mRNA that directs protein synthesis in the cytoplasm

Post-translational modification Modifications made to a polypeptide molecule after its initial synthesis, this includes proteolytic cleavages, *phosphorylation, glycosylation*, carboxylation, addition of fatty acid moieties, etc

Pre-mRNA An RNA molecule that is *transcribed* from chromosomal DNA in the nucleus of eukaryotic cells, and subsequently processed through *splicing* reactions to generate the mRNA that directs protein synthesis in the cytoplasm

Primary structure Refers to the sequence of *amino acid residues* or *nucleotides* within protein or nucleic acid molecules, respectively (also see SECONDARY and TERTIARY STRUCTURE)

Primer An *oligonucleotide* that is *complementary* to a specific region within a DNA or RNA molecule, and that is used to prime (initiate) synthesis of a new strand of complementary DNA at that specific site, in a reaction or series of reactions catalyzed by a *DNA polymerase*. The newly synthesized DNA strand will contain the primer at its 5′ end. Typically, primers are chemically synthesized oligonucleotides 15–50 nucleotides in length, selected on the basis of a known sequence (also known as 15–50 mers primers). However, "random primers" (shorter oligonucleotides, about 6 nucleotides in length, and comprising all possible sequences) may be used to prime DNA synthesis from DNA or RNA of unknown sequence. but probably serves to enhance stability of the RNA. Is frequently used to select mRNA for cloning purposes by *annealing* to a column containing a matrix bound to poly-uridylic acid

Primer-dimer At room temperature temporary association of primers may take place allowing the *Taq* enzyme to polymerise small strands of DNA called primer-dimers. Can give a false positive result if the detection of amplicons is via labelled primers

Primer extension A reaction in which DNA is *reverse transcribed* from an RNA *template* to which a specific oligonucleotide *primer* has been *annealed*. The new cDNA product is an extension of the primer, that is synthesized at the 3′ end of the primer in a direction extending toward the 5′ end of the RNA. This reaction is useful for exploring the extreme 5′ end of RNA molecules

Probe Usually refers to a DNA or RNA molecule that has been labeled with 32P or with *biotin*, to facilitate its detection after it has specifically *hybridized* with a target DNA or RNA sequence. However, the term may also refer to antibody probes used in *western* blots

Processing With respect to proteins, generally used to refer to proteolytic *post-translational modifications* of a polypeptide. In the case of RNA, processing may involve the addition of a 5′ *cap* and 3′ *poly-A* tracks as well as *splicing* reactions in the nucleus

Processivity The extent to which an RNA or DNA *polymerase* adheres to a *template* before dissociating, determines the average length (in kilobases) of the newly synthesized nucleic acid strands. Also applies to the action of *exonucleases* in digesting from the ends to the middle of a nucleic acid

Promoter A specific sequence within a double-stranded DNA molecule that is recognized by an RNA *polymerase*, that binds to it and uses it to begin transcribing the DNA *template* into a new RNA. The location and orientation of the *promoter* within a DNA molecule determines the start site of the new RNA. Other proteins (e.g. transcriptional activators such as *sigma factor*) are usually required for an RNA polymerase to recognize a promoter (see TRANSCRIPTION)

Proto-oncogene A cellular *oncogene*-like sequence that is thought to play a role in controlling normal cellular growth and differentiation

Pseudogene Inactive but stable components of the genome that derived by duplication and mutation of an ancestral, active gene. Pseudogenes can serve as the donor sequence in *gene conversion* events

Pseudorevertant A mutant virus or organism that has recovered a wildtype phenotype due to a second-site mutation (potentially located in a different region of the genome, or involving a different polypeptide) that has eliminated the effect of the initial mutation

Pseudoknot A feature of RNA *tertiary structure*; best visualized as two overlapping *stem-loops* in that the loop of the first stem-loop participates as half of the stem in the second stem-loop

Purine bases Adenine (A) or Guanine (G) (see NUCLEOTIDE)

Pulsed-field gel electrophoresis (PFGE) Separation of large (>50 kb) pieces of DNA, including complete chromosomes and genomes, by rapidly alternating the direction of electrophoretic migration in agarose gels

Pyrimidine bases Cytosine (C), Thymine (T) or Uracil (U) (see NUCLEOTIDE)

Q

Q-beta replicase An isothermal nucleic acid amplification system that uses the enzyme Q-beta replicase to replicate an RNA probe. An automated diagnostic system is being developed by Vysis Inc. (USA)

R

R coefficient Correlation coefficient, which is used to analyze a standard curve (ten-fold dilutions plotted against C_t values) obtained by linear regression analysis. It should be ≥ 0.99 for gene quantitation analysis. It takes values between zero and –1 for negative correlation and between zero and +1 for positive correlations

R^2 coefficient (coefficient of determination) This coefficient only takes values between zero and +1. R^2 is used to assess the fit of the standard curve to the data points plotted. The closer the value to 1, the better the fit

RACE Rapid Amplification of Complementary Ends

Reading frame Refers to a polypeptide sequence potentially encoded by a single-stranded *mRNA*. Because *codons* are nucleotide triplets, each mRNA has 3 reading frames (each nucleotide can participate in 3 codons, at the 1st, 2nd, and 3rd base position). Duplex DNA strands have 6 reading frames, 3 in each strand (see OPEN READING FRAME): AlaSerProLeuVal. 1st reading frame, ProAlaProTERTrp 2nd reading frame: TER = Stop, GlnProProSerGly 3rd reading frame

Real time PCR Real-time PCR (also known as Quantitative PCR, Real-time Quantitative PCR, or http://www.everythingbio.com/glos/definition.php? ID= 4079RTQ-PCR – but not **RT**-PCR) is a method of simultaneous DNA quantification and amplification. DNA is specifically amplified by polymerase chain reaction (PCR). After each round of amplification, the DNA is quantified. Real-time PCR can be combined with reverse transcription-polymerase chain reaction (http://www.everythingbio.com/glos/definition.php?ID=3471RT-PCR) to quantify low abundance messenger RNA, enabling a researcher to quantify relative gene expression at a particular time, or in a particular cell or tissue type. The combined technique is often called quantitative http://www.everythingbio.com/glos/definition.php?ID=3471RT-PCR

RT-PCR (Reverse transcriptase polymerase chain reaction) A variation of the PCR technique in which cDNA is made from RNA via reverse transcription. The cDNA is then amplified using standard PCR protocols

Recognition sequence A specific *palindromic sequence* within a double-stranded DNA molecule that is recognized by a *restriction endonuclease*, and at which the restriction endonuclease specifically cleaves the DNA molecule

Recombination-repair A mode of filling a gap in one strand of duplex DNA by retrieving a homologous single strand from another duplex. Usually the underlying mechanism behind *homologous recombination* and *gene conversion*

Relative quantitation Method in which unknown samples are compared to reference samples to determine the increase or decrease of gene expression. The out come is presented as an index, not an absolute amount

Relaxed DNA See SUPERCOIL

Replication The copying of a nucleic acid molecule into a new nucleic acid molecule of similar type (i.e., DNA > DNA, or RNA > RNA)

Reporter dye Fluorescent dye used to monitor the PCR product accumulation. This can be attached to a specific probe or can be a dsDNA-binding agent (see for example **SYBR® Green I**)

Reporter gene The use of a functional enzyme, such as beta-galactosidase, luciferase, or chloramphenicol acetyltransderase, downstream of a gene, promoter, or translational control element of interest, to more easily identify successful introduction of the gene into a host and to measure transcription and/or translation

Repression Inhibition of transcription (or translation) by the binding of a repressor protein to a specific site on DNA (or *mRNA*

Residue As applied to proteins, what remains of an *amino acid* after its incorporation into a peptide chain, with subsequent loss of a water molecule (see PEPTIDE BOND)

Restriction endonuclease (RE) A bacterial enzyme that recognizes a specific *palindromic sequence* (*recognition sequence*) within a double-stranded DNA molecule and then catalyzes the cleavage of both strands at that site. Also called a restriction enzyme. Restriction endonucleases may generate either *blunt* or *sticky ends* at the site of cleavage

Restriction fragment length polymorphism (RFLP) Variations in the lengths of fragments of DNA generated by digestion of different DNAs with a specific *restriction endonuclease*, reflecting genetic variation (*polymorphism*) in the DNAs

Restriction fragments DNA fragments generated by digestion of a DNA preparation with one or more *restriction endonucleases*; usually separated by *agarose gel electrophoresis* and visualized by *ethidium bromide* staining under UV light (or alternatively subjected to *Southern blot analysis*)

Restriction map A linear array of sites on a particular DNA that are cleaved by various selected *restriction endonucleases*

Reticulocyte lysate A lysate of rabbit reticulocytes, that has been extensively digested with micrococcal nuclease to destroy the reticulocyte *mRNA*s. With the addition of an exogenous, usually synthetic, mRNA, *amino acids* and a source of energy (ATP), the translational machinery of the reticulocyte (*ribosomes*, eukaryotic translation factors, etc.) will permit in vitro *translation* of the added mRNA with production of a new *polypeptide*. This is only one of several available in vitro translation systems

Reverse transcriptase (RT) A *DNA polymerase* that copies an RNA molecule into single-stranded cDNA; usually purified from retroviruses

Reverse transcription Copying of an RNA molecule into a DNA molecule

RT-PCR Real time PCR

Ribonuclease (RNASE) An enzyme that catalyzes the hydrolysis of RNA. There are many different RNases, some of the more important include

RNASE H Degrades the RNA part of RNA:DNA hybrids

Ribosomal binding sequence (shine-dalgarno sequence) In prokaryotic organisms, part or all of the polypurine sequence AGGAGG located on *mRNA* just upstream of an AUG *initiation codon*; it is complementary to the sequence at the 3′ end of 16S rRNA; and involved in binding of the ribosome to *mRNA*. The *internal ribosomal entry site* found in some viruses may be an analogous eukaryotic genetic element

Ribosome A complex ribonucleoprotein particle (eukaryotic ribosomes contain 4 RNAs and at least 82 proteins) which is the "machine" which translates *mRNA* into protein molecules. In eukaryotic cells, ribosomes are often in close proximity to the endoplasmic reticulum

Ribozyme A catalytically active RNA. A good example is the hepatitis delta virus RNA that is capable of self-cleavage and self-ligation in the absence of protein enzymes

Rn Also known as Normalized reporter signal. The fluorescence emission intensity of the reporter dye divided by the fluorescence emission intensity of the passive reference dye

RNA polymerase A polymerase that synthesizes RNA (see POLYMERASE)

RNA splicing A complex and incompletely understood series of reactions occurring in the nucleus of eukaryotic cells in which *pre-mRNA transcribed* from chromosomal DNA is processed such that noncoding regions of the pre-mRNA (*introns*) are excised, and coding regions (*exons*) are covalently linked to produce an *mRNA* molecule ready for transport to the cytoplasm. Because of splicing, eukaryotic DNA representing a gene encoding any given protein is usually much larger than the mRNA from which the protein is actually *translated*

RNASE see RIBONUCLEASE

rRNA Ribosomal RNA (four sizes in humans: 5S, 5.8S, 18S, and 28S); RNA component of the *ribosome* that may play catalytic roles in *translation*

ROX 6-carboxy-X-rhodamine. Most commonly passive reference dye for normalization of reporter signal

RT Reverse Transcriptase

RT/PCR reaction A series of reactions that result in RNA being copied into DNA and then amplified. A single *primer* is used to make single-stranded *cDNA* copies from an RNA *template* under direction of *reverse transcriptase*. A second primer *complementary* to this "first strand" cDNA is added to the reaction mix along with

Taq polymerase, resulting in synthesis of double-stranded DNA. The reaction mix is then cycled (denaturation, *annealing* of primers, *extension*) to amplify the DNA by conventional PCR

Runoff transcript RNA that has been synthesized from plasmid DNA (usually by a RNase A Cleaves ssRNA 3′ of pyrimidines, RNase T1 Cleaves ssRNA at guanasine nucleotides, RNase V1 Cleaves dsRNA (helical regions)bacteriophage *RNA polymerase* such as *T7* or *SP6*) and which terminates at a specific 3′ site because of prior cleavage of the plasmid DNA with a *restriction endonuclease*.
RNA. Ribonucleic acid. Single strand nucleic acid composed of a phosphate ribose "backbone" and the nucleotides: adenine, guanine, cytosine and uracil

mRNA Messenger RNA, carries genetic information from DNA to the ribosome where it is translated into the corresponding protein

S

Scorpion Fluorescence detection system that consists of a detection probe with the upstream primer with a fluorophore at the 5′ end, followed by a complementary stem-loop structure also containing the specific probe sequence, quencher dye and a PCR primer on the 3′ end. Between the primer and its tail (the probe), a blocking agent is placed. Usually used for its specificity on allelic discrimination assays

Semi or hemi-nested PCR If a full nested PCR (using two internal primers) cannot be performed, sensitivity and specificity can be improved by using just one "inner" primer in conjunction with one of the "outer" primers from the first reaction

S1 nuclease An enzyme that digests single-stranded DNA or RNA

SDS-page Denaturing protein gel electrophoresis (see POLYACRYLAMIDE GEL ELECTROPHORESIS)

Selection The use of particular conditions, such as the presence of ampicillin, to allow survival only of cells with a particular *phenotype*, such as production of beta-lactamase

Sequence polymorphism See Polymorphism

Shotgun cloning or sequencing Cloning of an entire genome or large piece of DNA in the form of randomly generated small fragments. The individual sequences obtained from the clones will be used to construct *contigs*

Shuttle vector A small *plasmid* capable of *transfection* into both prokaryotic and eukaryotic cells

Sigma factor Certain small ancillary proteins in bacteria that increase the binding affinity of RNA polymerase to a promoter. Different sigma factors recognize different promoter sequences

Signal peptidase An enzyme present within the lumen of the endoplasmic reticulum that proteolytically cleaves a secreted protein at the site of a *signal sequence*

Signal sequence A hydrophobic amino acid sequence that directs a growing peptide chain to be secreted into the endoplasmic reticulum

Silent mutation A nucleotide substitution (never a single deletion or insertion) that does not alter the amino acid sequence of an encoded protein due to the *degeneracy* of the genetic code. Such mutations usually involve the third base (*wobble* position) of *codons*

Site-directed mutagenesis The introduction of a mutation, usually a *point mutation* or an insertion, into a particular location in a cloned DNA fragment. This mutated fragment may be used to "*knock out*" a gene in the organism of interest by *homologous recombination*

Site-specific recombination Occurs between two specific but not necessarily homologous sequences. Usually catalyzed by enzymes not involved in general or *homologous recombination*

SOP Standard Operating procedure

Southern blot DNA is separated by electrophoresis (usually in *agarose gels*), then transferred to nitrocellulose paper or other suitable solid-phase matrix (e.g., nylon membrane), and denatured into single strands so that it can be *hybridized* with a specific *probe*. The Southern blot was developed by E.M. Southern, a molecular biologist in Edinburgh. *Northern* and *western* blots were given contrasting names to reflect the different target substances (RNA and proteins, respectively) that are subjected in these procedures to electrophoresis, blotting and subsequent detection with specific probes

Southwestern blot The binding of protein to a nucleic acid on a matrix similar to what is done for western, northern, and southern blots. This technique is used to identify DNA binding proteins and the recognition sites for these proteins

SP6 RNA polymerase A bacteriophage *RNA polymerase* that is commonly used to transcribe *plasmid* DNA into RNA. The plasmid must contain an SP6 *promoter* upstream of the relevant sequence

Splicing see RNA SPLICING

ss Single stranded

ss DNA Single stranded DNA

ss RNA Single stranded RNA

Standard A sample of known concentration used to construct a standard curve. By running standards of varying concentrations, a standard curve is created from which the quantity of an unknown sample can be calculated

Standard curve Plotting of C_t values against log-transformed concentrations of serial ten-fold dilutions of the target nucleic acid. It is used to find out the dynamic range of the target (and/or normalizer), to calculate the slope (therefore, efficiency), r and R^2 coefficients and also to help with quantitation

Start codon See Initiation codon

Stem-loop A feature of RNA *secondary structure*, in which two complementary, inverted sequences that are separated by a short-intervening sequence within a single strand of RNA base pair to form a "stem" with a "loop" at one end. Similar to a *hairpin*, but these usually have very small loops and longer stems

Sticky end Terminus of DNA molecule that has either a 3′ or 5′ overhang, and which typically results from a cut by a *restriction endonuclease*. Such termini are capable of specific ligation reactions with other termini that with complementary overhangs. A sticky end can be "blunt ended" by the removal of an overhang, or a "filling in" reaction that adds additional nucleotides complementary to overhang (see BLUNT END)

SDA Strand Displacement Assay. An isothermal nucleic acid amplification system where cDNA product is synthesised from an RNA target. Covered by patents owned by Becton Dickinson (USA)

SYBR® green I Fluorogenic minor groove binding dye that emits no fluorescence in solution, but emits strong fluorescent signal upon binding to double-stranded DNA. It does not bind to ssDNA but binds to any dsDNA product

T

Ta Annealing temperature

TAE Tris acetic acid EDTA

TAMRA 6-carboxy-terta-methyl-rhodamine. Most commonly used quencher at the 3′ end of a TaqMan® probe

Target The specific piece of DNA or RNA to be amplified by the PCR. The sensitivity of an assay can be enhanced if multiple copies of the target are present. e.g. a repeated sequence or mRNA molecules

TAQ Thermostable DNA polymerase, originally isolated from *Thermus aquaticus,* an organism that resides in hot springs

TaqMan assay 5′nuclease fluorogenic assay. For simple and rapid detection of PCR products. An oligonucleotide probe that is specific for the target to be amplified is labelled with a fluorescent tag and a quenching molecule. During the extension step of a PCR the Taq enzyme will disrupt probe bound to the target separating the fluorescent tag from its quencher molecule thus permitting fluorescence

Taq polymerase A *DNA polymerase* that is very stable at high temperatures, isolated from the thermophilic bacterium *Thermus aquaticus*. Very useful in *PCR*

reactions which must cycle repetitively through high temperatures during the denaturation step

TBE Tris boric acid EDTA

Template A nucleic acid strand, upon which a *primer* has *annealed* and a nascent RNA strand is being extended

Termination codon See stop codon

Terminator A sequence *downstream* from the 3′ end of an *open reading frame* that serves to halt *transcription* by the RNA polymerase. In bacteria these are commonly sequences that are *palindromic* and thus capable of forming *hairpins*. Sometimes termination requires the action of a protein, such as Rho factor in *E. coli*

Thermal cycler Although PCR may be performed using manual transfer between a series of water or oil baths, for reproducible results or diagnostic work a reliable programmable thermal cycling machine is essential. There are now many machines to choose from with a wide range of specifications and costs. A patent on the use of thermal cycling machines for PCR is held by the Perkin Elmer Corporation (USA). It should be remembered that heating and cooling parameters vary with the design of the instrument and a PCR that has been optimised for one machine may not function as efficiently on a different type of thermal cycler

Three room strategy In order to minimise the risks of contamination 3 workstations should be used, each in a separate room. A "Clean" room for preparation of PCR reagents, a "Grey" area for preparation of samples and positive controls and a "Dirty" room for post-amplification work and detection of amplicons

Threshold Usually 10X the standard deviation of Rn for the early PCR cycles (baseline). The threshold should be set in the region associated with an exponential growth of PCR product (which may be easier is the log-view of the amplification plot is used). It is assigned for each run to calculate the C_t value for each amplification

Tm The midpoint of the temperature range over which DNA is melted or denatured by heat; the temperature at that a *duplex* nucleic acid molecule is 50% *melted* into single strands, it is dependent upon the number and proportion of G-C *base pairs* as well as the ionic conditions. Often referred to as a measure of the thermal stability of a nucleic acid *probe*:target sequence hybrid. To calculate the approximate Tm for an oligonucleotide count the numbers of each nucleotide $Tm = 2(A+T) + 4\ (G + C)$

TMA Transcription-mediated amplification, an isothermal nucleic acid amplification system. Amplification of RNA using RNA polymerase. Covered by patents owned by Gen-Probe Incorporated (USA). A range of diagnostic kits is available

T7 RNA polymerase A bacteriophage *RNA polymerase* that is commonly used to transcribe plasmid DNA into RNA. The plasmid must contain a T7 *promoter* upstream of the relevant sequence

Trans As used in molecular biology, an interaction that involves two sites that are located on separate molecules

Transcript A newly made RNA molecule that has been copied from DNA

Transcription The copying of a DNA template into a single-stranded RNA molecule. The processes whereby the transcriptional activity of eukaryotic genes are regulated are complex, involve a variety of accessory transcriptional factors that interact with *promoters* and *polymerases*, and constitute one of the most important areas of biological research today

Transcription/translation reaction An in vitro series of reactions, involving the synthesis (*transcription*) of an mRNA from a *plasmid* (usually with *T7* or *SP6 RNA polymerase*), followed by use of the mRNA to program *translation* in a cell-free system such as a rabbit *reticulocyte lysate*. The *polypeptide* product of translation in usually labelled with [35S]-methionine, and examined in an *SDS-PAGE* gel with or without prior *immunoprecipitation*. This series of reactions permits the synthesis of a polypeptide from DNA in vitro

Transcriptional start site The nucleotide of a gene or cistron at which *transcription* (RNA synthesis) starts; the most common triplet at which transcription begins in *E. coli* is CAT. *Primer extension* identifies the transcriptional start site

Transfection The process of introducing foreign DNA (or RNA) into a host organism, usually a eukaryotic cell

Transformation Multiple meanings. With respect to cloning of DNA, refers to the transformation of bacteria (usually to specific antibiotic resistance) due to the uptake of foreign DNA. With respect to eukaryotic cells, usually means conversion to less-restrained or unrestrained growth

Transgene A foreign gene that has been introduced into the germ line of an animal species

Transgenic An animal (usually a mouse) or plant into which a foreign gene has been introduced in the germ line. An example: transgenic mice expressing the human receptor for poliovirus are susceptible to human polioviruses

Transition A nucleotide substitution in which one pyrimidine is replaced by the other pyrimidine, or one purine replaced by the other purine (e.g., A is changed to G, or C is changed to T) (contrast with TRANSVERSION)

Translation The process whereby *mRNA* directs the synthesis of a protein molecule; carried out by the *ribosome* in association with a host of translation initiation, elongation and termination factors. Eukaryotic genes may be regulated at the level of translation, as well as the level of *transcription*

Translocation The process by which a newly synthesized protein is directed toward a specific cellular compartment (i.e., the nucleus, the endoplasmic reticulum)

Transposon A transposable genetic element; certain sequence elements that are capable of moving from one site to another in a DNA molecule without any requirement for sequence relatedness at the donor and acceptor sites. Many transposons carry antibiotic resistance determinants and have insertion sequences at both ends, and thus have two sets of inverted repeats

Transposition The movement of DNA from one location to another location on the same molecule, or a different molecule within a cell

Transversion A nucleotide substitution in which a purine replaces a pyrimidine, or vice versa (e.g., A is changed to T, or T is changed to G) (see TRANSITION)

Triplet A three-nucleotide sequence; a *codon*

tRNA Small, tightly folded RNA molecules which act to bring specific amino acids into *translationally* active *ribosomes* in a fashion that is dependent upon the *mRNA* sequence. One end of the tRNA molecule recognizes the nucleotide triplet which is the *codon* of the mRNA, while the other end (when activated) is covalently linked to the relevant *amino acid*

U

UNG Uracil N-glycosylase

Unknown Experimental sample containing an unknown quantity of template

Untranslated RNA See Nontranslated RNA

Upstream Identifies sequences located in a direction opposite to that of expression; for example, the bacterial *promoter* is upstream of the *initiation codon*. In an mRNA molecule, upstream means toward the 5′ end of the molecule. Occasionally used to refer to a region of a polypeptide chain that is located toward the amino terminus of the molecule

UV Ultra violet

Index

MIX
Papier aus verantwortungsvollen Quellen
Paper from responsible sources
FSC® C105338

FSC
www.fsc.org

If you have any concerns about our products,
you can contact us on
ProductSafety@springernature.com

In case Publisher is established outside the EU,
the EU authorized representative is:
Springer Nature Customer Service Center GmbH
Europaplatz 3, 69115 Heidelberg, Germany

Printed by Libri Plureos GmbH
in Hamburg, Germany